# Chemolithoautotrophic Bacteria

## Biochemistry and Environmental Biology

Tateo Yamanaka

# Chemolithoautotrophic Bacteria

## Biochemistry and Environmental Biology

With 47 Figures

Springer

Tateo Yamanaka, Ph.D.
Professor Emeritus
Tokyo Institute of Technology
1470 Iwameji, Hidaka-mura
Kochi 781-2154, Japan

Library of Congress Control Number: 2008922900

ISBN 978-4-431-78540-8 Springer Tokyo Berlin Heidelberg New York
e-ISBN 978-4-431-78541-5



Springer is a part of Springer Science+Business Media
springer.com


Printed in Japan

Typesetting: SNP Best-set Typesetter Ltd., Hong Kong
Printing and Binding: Kato Bunmeisha, Japan

Printed on acid-free paper

# Preface

The readers of this book probably will not easily believe it when I say that wooden houses of one or two stories can be destroyed by the action of bacteria residing in the ground under the floor. Of course, the force of one bacterium cell is very, very weak. However, when $10^{14}$ cells/$m^3$ each of sulfate-reducing, sulfur-oxidizing, and acidophilic iron-oxidizing bacteria reside in the ground under the houses, they will produce a great amount of sulfuric acid, and the sulfuric acid thus formed reacts with calcium carbonate in the soil to form gypsum under certain conditions. Crystallization of gypsum makes the soil bulky, and the ground heterogeneously heaves and damages the houses. The maximal height of the heaving seen so far is 48 cm. As a result, pillars of houses are tilted and beams are pushed upward as part of the heaving of the foundations.

Previously, no one had believed that concrete was corroded by bacteria. It is generally accepted, however, that the surface of the concrete used for sewerage systems is sometimes so corroded by bacteria that they can be destroyed by scratching with a spatula. The cause of this corrosion has been clarified: hydrogen sulfide is formed by the action of the sulfate-reducing bacteria in the sewage, it rises to the atmosphere in the sewerage systems, is oxidized by the sulfur-oxidizing bacteria to become sulfuric acid on the surface of concrete, and the resulting sulfuric acid corrodes the concrete. The phenomenon is one expression or stage of the circulation of sulfur on Earth.

Bacteria also participate in the circulation of nitrogen on Earth. Nitrogen gas changes form, becoming ammonia, then nitrite, then nitrate, and finally returning to being nitrogen gas, and special bacteria participate in each step of the changes. Acidophilic iron-oxidizing bacteria dissolve metals, especially in metal mines. It could thus be said that bacteria change the surface of the Earth. The bacteria are applied in collecting metals from metal ores (bacterial leaching) and in cleansing the mine sewage.

In the phenomena mentioned above, the bacteria that participate intimately grow only on inorganic compounds. They are referred to as chemolithoautotrophic bacteria. My colleagues and I have actively studied the biochemistry and physiology of chemolithoautotrophic bacteria, and in this book I want to summarize the work done to date on several of the chemolithoautotrophic bacteria and on their relationships to the environment.

In the biosphere of the Earth, i.e., in soil, water, and air, various kinds of bacteria reside. Although they sometimes cause such damage as that mentioned above, many of them, especially the chemolithoautotrophic bacteria, help us by participating in the circulation of various substances on Earth. Microorganisms including bacteria also decompose organic trash and garbage and cleanse the surface of the Earth. Although, of course, the decomposition of organic substances by heterotrophic bacteria is very important in the sense of cleansing the surface of the Earth, in this book I particularly want to describe the relationship to the environment of the chemolithoautotrophic bacteria that participate in changing inorganic substances on Earth, in relation to the work my colleagues and I have done. I also want to emphasize the fact that human beings interact with the Earth through chemolithoautotrophic bacteria.

Chapter 1 treats the growth of the microorganisms: how they eat what kinds of foods or feed to grow. The mechanisms by which the organisms of various nutritional types grow are described schematically.

In Chapter 2, there is a brief survey of cytochromes, which participate especially in the bacterial oxidation of inorganic compounds.

Chapter 3 deals with the circulation of nitrogen on Earth. The main bacteria that participate in nitrogen circulation are ammonia-oxidizing bacteria, nitrite-oxidizing bacteria, denitrifying bacteria, and nitrogen-fixing bacteria. The bacterial oxidation mechanisms of ammonia and nitrite are described in some detail. As the applications of the nitrifying bacteria (i.e., ammonia-oxidizing and nitrite-oxidizing bacteria), the bacterial production of gunpowder, and the removal of ammonia from sewage are described. An incident related to bacterial nitrification is also mentioned, and the formation of nitric oxide in human tissues and some of its physiological functions are described.

In Chapter 4, the circulation of sulfur on Earth is described, including the reduction of sulfate by the sulfate-reducing bacteria and the oxidation of sulfur compounds by the sulfur-oxidizing bacteria. First, the reduction mechanisms of sulfate by the sulfate-reducing bacteria are mentioned. In relation to the bacteria, there is presented the possibility of determining how long ago life originated. Next, the oxidation mechanisms of sulfur compounds by the sulfur-oxidizing bacteria are described, including the fact that sulfur-oxidizing bacteria support animal life around hydrothermal vents on the dark deep-sea bottom. This chapter also describes some of the damage—specifically, the corrosion of concrete—caused by sulfuric acid when hydrogen sulfide is oxidized by the bacteria on concrete surfaces.

Chapter 5 deals with the bacterial oxidation and reduction of iron. First, the oxidation mechanisms of iron by bacteria are explained. Next, applications of bacteria are described: bacterial leaching, biohydrometallurgy, and other processes. Also described in this chapter is the heaving of house foundations caused by the "cooperation" of sulfate-reducing bacteria, sulfur-oxidizing bacteria, and acidophilic iron-oxidizing bacteria. The heaving results in damage especially to wooden structures.

In Chapter 6, the circulation of carbon on Earth is described. Although the consumption of carbon dioxide by plants, algae, cyanobacteria, and anoxygenic photosynthetic bacteria is a large factor in carbon circulation, it is treated only briefly here because the phenomenon is a major topic in discussions of photosynthesis. However, there is a fairly detailed description of the pathways by which organic compounds are formed from carbon dioxide, because chemolithoautotrophic organisms must form cellular materials from carbon dioxide. Although a description of photosynthesis is omitted, the bacterial formation of methane is described in some detail, with the explanation that methane formation by methanogens is not by fermentation but, rather, through respiration. Additionally, there is a description of methods to reduce the evolution of methane from rice paddies in Japan.

Finally, in Chapter 7, consideration is given to the bacteria that are thought to be evolutionally situated most closely to the origin of life. Hyperthermophilic bacteria are considered to be the organisms closest to the origin of life, especially on the basis of studies of RNA sequences. Many hyperthermophilic bacteria are anaerobic chemolithoautotrophs. Discussed here are the energy-acquiring systems of the organisms that are thought to have been present at earlier evolutional stages. As I make clear, I am convinced that neither the Embden–Meyerhof–Parnas pathway nor the Entner–Doudoroff pathway is the energy-acquiring system of the evolutionally earliest organisms.

My work has been involved with cytochromes of various organisms and in the biochemistry and physiology of several chemolithoautotrophic bacteria. For that reason, Chapters 2, 3, 4, and 5 are written mainly on the basis of my own work. Although my contribution to the studies included in Chapter 6 is very limited, I have purified cytochrome *bc* from a methanogen. Also, because I have studied molecular evolution based on studies of cytochromes, Chapter 7 presents a strong statement of my ideas about the energy-acquiring processes in organisms at earlier evolutional stages.

I am very grateful to all my colleagues whose names appear with mine in the literature cited. I thank Professor Yoko Nagata and Mr. Minoru Tanigawa (College of Science and Technology, Nihon University) for their assistance in checking references and for technical assistance in completing this manuscript, respectively, and I express my appreciation to the editorial staff members of Springer Japan for their help in publishing this volume.

Tateo Yamanaka
Professor Emeritus
Tokyo Institute of Technology
*January 2008*

# Contents


**Preface** . . . . . . . . . . . . . . . . . . . . . . . . . . . . . . . . . V

**Abbreviations** . . . . . . . . . . . . . . . . . . . . . . . . . . . . . XIII

**1 General Considerations** . . . . . . . . . . . . . . . . . . . . . . . . 1
1.1 Chemoheterotrophic Bacteria . . . . . . . . . . . . . . . . . . . . 2
1.2 Chemolithoautotrophic Bacteria . . . . . . . . . . . . . . . . . . 4
1.2.1 Ammonia-Oxidizing Bacteria . . . . . . . . . . . . . . . . 5
1.2.2 Nitrite-Oxidizing Bacteria . . . . . . . . . . . . . . . . . 5
1.2.3 Denitrifying Bacteria . . . . . . . . . . . . . . . . . . . . 6
1.2.4 Sulfate-Reducing Bacteria . . . . . . . . . . . . . . . . . 7
1.2.5 Sulfur-Oxidizing Bacteria . . . . . . . . . . . . . . . . . 8
1.2.6 Iron-Oxidizing and -Reducing Bacteria . . . . . . . . . . . 8
1.2.7 Methanogens . . . . . . . . . . . . . . . . . . . . . . . . 9

**2 Cytochromes** . . . . . . . . . . . . . . . . . . . . . . . . . . . . . 11
2.1 Hemes . . . . . . . . . . . . . . . . . . . . . . . . . . . . . . . 11
2.2 Kinds of Cytochromes . . . . . . . . . . . . . . . . . . . . . . . 11
2.2.1 Heme A-Containing Cytochromes . . . . . . . . . . . . . 11
2.2.2 Heme B-Containing Cytochromes . . . . . . . . . . . . . 14
2.2.3 Heme C-Containing Cytochromes . . . . . . . . . . . . . 15
2.2.4 Heme $D_1$-Containing Cytochromes . . . . . . . . . . . . 15
2.2.5 Heme O-Containing Cytochromes . . . . . . . . . . . . . 16
2.2.6 Heme D-Containing Cytochromes . . . . . . . . . . . . . 16

**3 Nitrogen Circulation on Earth and Bacteria** . . . . . . . . . . . . . 17
3.1 Bacterial Nitrification . . . . . . . . . . . . . . . . . . . . . . . 18
3.1.1 Oxidation of Ammonia . . . . . . . . . . . . . . . . . . . 18
3.1.2 Oxidation of Hydroxylamine . . . . . . . . . . . . . . . . 20
(a) Hydroxylamine Oxidoreductase . . . . . . . . . . . . . 20
(b) Cytochrome *c*-554 . . . . . . . . . . . . . . . . . . . 24

(c) Cytochrome *c*-552 . . . . . . . . . . . . . . . . . . . 25
(d) Cytochrome *c* Oxidase . . . . . . . . . . . . . . . . . . 25
3.1.3 Electron Transfer Pathway Coupled to the Oxidation of Ammonia . . . . . . . . . . . . . . . . . 27
3.1.4 Dehalogenation of Chloroethylenes by Bacteria . . . . . . 28
3.1.5 Various Growth Features of Ammonia-Oxidizing Bacteria . . . . . . . . . . . . . . . . . 30
3.1.6 Bacterial Oxidation of Nitrite . . . . . . . . . . . . . . . . . 31
(a) Nitrite Oxidoreductase . . . . . . . . . . . . . . . . . . 31
(b) Cytochromes *c*-550(s) and *c*-550(m) . . . . . . . . . . 34
(c) Cytochrome *c* Oxidase . . . . . . . . . . . . . . . . . . 34
(d) Reconstitution of Nitrite Oxidation System . . . . . . 35
3.1.7 Nitrification by Heterotrophic Bacteria . . . . . . . . . . . 37
3.2 Applications of Nitrifying Bacteria . . . . . . . . . . . . . . . . . 38
3.2.1 Bacterial Production of Gunpowder . . . . . . . . . . . . . 38
3.2.2 Removal of Ammonia from Sewage . . . . . . . . . . . . . 39
3.3 Interaction Between Ammonia-Oxidizing and Nitrite-Oxidizing Bacteria . . . . . . . . . . . . . . . . . . . . . . . . 40
3.3.1 Was Earth Previously Polluted by Nitrite? . . . . . . . . . . 40
3.3.2 An Agricultural Incident Caused by Incomplete Nitrification . . . . . . . . . . . . . . . . . . . 41
3.3.3 Herbicides and Nitrification . . . . . . . . . . . . . . . . . . 42
3.4 Reduction of Nitrate and Nitrogen Gas . . . . . . . . . . . . . . 44
3.4.1 Bacteria That Reduce Nitrate to Nitrogen Gas . . . . . . . 44
3.4.2 Nitric Oxide Is also Produced in Human Tissues . . . . . . 46
3.4.3 Bacteria Reducing Nitrogen Gas to Ammonia . . . . . . . 47
(a) Rhizobia . . . . . . . . . . . . . . . . . . . . . . . . . . 49
(b) Azotobacter . . . . . . . . . . . . . . . . . . . . . . . . . 50
(c) Cyanobacteria . . . . . . . . . . . . . . . . . . . . . . . 51

**4 Sulfur Circulation on Earth and Bacteria . . . . . . . . . . . . . . . . 53**
4.1 Bacteria Forming Hydrogen Sulfide . . . . . . . . . . . . . . . . 53
4.1.1 Bacterial Reduction Mechanisms of Sulfate . . . . . . . . . 54
4.1.2 Components Participating in Bacterial Reduction of Sulfate . . . . . . . . . . . . . . . . . . . . . . 56
(a) Hydrogenase . . . . . . . . . . . . . . . . . . . . . . . . 56
(b) Cytochromes . . . . . . . . . . . . . . . . . . . . . . . . 56
(c) Adenylylsulfate Reductase . . . . . . . . . . . . . . . . 58
(d) Sulfite Reductase . . . . . . . . . . . . . . . . . . . . . . 58
(e) Siroheme . . . . . . . . . . . . . . . . . . . . . . . . . . 59
4.1.3 Sulfate-Reducing Bacteria and Molecular Oxygen . . . . . . . . . . . . . . . . . . . . . . . . 61
4.1.4 Sulfur Respiration . . . . . . . . . . . . . . . . . . . . . . . 61
4.1.5 Autumnal Dying of Rice Plants . . . . . . . . . . . . . . . . 62
4.1.6 Checking How Old the Origin of Life Is . . . . . . . . . . . 63

4.2 Sulfur-Oxidizing Bacteria . . . . . . . . . . . . . . . . . . . . 64
4.2.1 Bacterial Oxidation Mechanisms of Sulfur Compounds . . . 65
(a) Oxidation of Sulfide and Elemental Sulfur . . . . . . . 65
(b) Oxidation of Thiosulfate . . . . . . . . . . . . . . . . . 66
(c) Oxidation of Sulfite . . . . . . . . . . . . . . . . . . . 67
(d) Cytochrome *c* . . . . . . . . . . . . . . . . . . . . . . 68
(e) Cytochrome *c* Oxidase . . . . . . . . . . . . . . . . . 69
(f) Oxidation Systems of Sulfite and Thiosulfate . . . . . 70
4.2.2 Sulfur-Oxidizing Bacteria Support Animals in the Dark on the Deep-Sea Bottom . . . . . . . . . . . . . . . . . 72
4.2.3 Bacterial Corrosion of Concrete . . . . . . . . . . . . . . 74

**5 Oxidation and Reduction of Iron by Bacteria . . . . . . . . . . . . 79**
5.1 Bacteria That Oxidize or Reduce Iron . . . . . . . . . . . . . . 79
5.1.1 Mechanisms in Bacterial Oxidation of Iron . . . . . . . . . 81
(a) Fe(II)-Cytochrome *c* Oxidoreductase . . . . . . . . . 81
(b) Cytochromes *c* . . . . . . . . . . . . . . . . . . . . . 82
(c) Rusticyanin . . . . . . . . . . . . . . . . . . . . . . . 83
(d) Cytochrome *c* Oxidase . . . . . . . . . . . . . . . . . 84
(e) Electron Transfer System Coupled to Oxidation of Ferrous Ion . . . . . . . . . . . . . . . . 85
5.1.2 Oxidation of Sulfur Compounds by Iron-Oxidizing Bacteria . . . . . . . . . . . . . . . . . . 88
5.1.3 Various Growth Aspects of Acidothiobacillus Ferrooxidans . . . . . . . . . . . . . . . . 89
5.1.4 Iron-Oxidizing Bacteria Requiring No Oxygen . . . . . . . 91
5.1.5 Bacterial Reduction of Ferric Compounds . . . . . . . . . . 92
5.1.6 Bacteria Containing Magnetism . . . . . . . . . . . . . . . 93
5.2 Applications of Iron-Oxidizing Bacteria . . . . . . . . . . . . . . 94
5.2.1 Bacterial Leaching . . . . . . . . . . . . . . . . . . . . . . 94
5.2.2 Etching of Copper Plate . . . . . . . . . . . . . . . . . . . 96
5.2.3 Concentration of Gold from Pyrite Containing a Trace of Gold . . . . . . . . . . . . . . . . . . . . . . . 96
5.2.4 Biohydrometallurgy . . . . . . . . . . . . . . . . . . . . . 97
5.2.5 Cleaning of Mine Sewage . . . . . . . . . . . . . . . . . . 97
5.3 Upheaval of House Foundations: Damage Caused by Bacteria . . . 99

**6 Carbon Circulation on Earth and Microorganisms . . . . . . . . . 103**
6.1 Mechanisms of Formation of Organic Compounds from Carbon Dioxide . . . . . . . . . . . . . . . . . . 105
6.1.1 Calvin–Benson Cycle (Reductive Pentose Phosphate Cycle) . . . . . . . . . . . . . . . . . . . . . . 105
6.1.2 Hatch–Slack Pathway . . . . . . . . . . . . . . . . . . . . 106
6.1.3 Carbon Dioxide-Fixing Pathways Other than the Calvin–Benson Cycle in the Lithoautotrophs . . . . . . . 108

6.2 Methanogens . . . . . . . . . . . . . . . . . . . . . . . . . . . . 109
6.2.1 Mechanism of Lithoautotrophic Methane Formation: Respiration but Not Fermentation . . . . . . . . . . . . . . 110
6.2.2 Formation of Methane from Acetate . . . . . . . . . . . . 113
6.2.3 Methanogens and Cytochromes . . . . . . . . . . . . . . . 114
6.2.4 Methanogens and the Environment . . . . . . . . . . . . . 115
6.3 Bacteria Utilizing Carbon Monoxide . . . . . . . . . . . . . . . . . 116

**7 Organisms Evolutionarily Closest to the Origin of Life . . . . . . . 119**
7.1 Archaea and Their Energy-Acquiring Reactions . . . . . . . . . . . 119
7.2 Biological Evolution at Earlier Stages . . . . . . . . . . . . . . . . 121

**References . . . . . . . . . . . . . . . . . . . . . . . . . . . . . . . . 125**

**Index . . . . . . . . . . . . . . . . . . . . . . . . . . . . . . . . . . 153**

# Abbreviations

| | |
|---|---|
| AMO | Ammonia monooxygenase |
| APS | Adenosine 5′-phosphosulfate or adenylyl sulfate |
| Cyt | Cytochrome |
| Cyt *c* | Cytochrome *c* |
| Cyt *c*($Fe^{3+}$) | Ferricytochrome *c* |
| Cyt *c*($Fe^{2+}$) | Ferrocytochrome *c* |
| CoBSH | Coenzyme B or 7-mercaptoheptanoylthreonine phosphate (HS-HTP) |
| CoMSH | Coenzyme M ($HSCH_2CH_2SO_3^-$) |
| DCCD | Dicyclohexylcarbodiimide |
| DCIP | 2,6-dichlorophenolindophenol |
| $E_{m,n}$ | Midpoint redox potential at pH *n*. For example, $E_{m,7.0}$ is midpoint redox potential at pH 7.0 |
| EPR | Electron paramagnetic resonance |
| $F_{420}$ | Coenzyme $F_{420}$ |
| $F_{430}$ | Coenzyme $F_{430}$ |
| GSSG | Oxidized form of glutathione |
| GSH | Reduced form of glutathione |
| HAO | Hydroxylamine oxidoreductase |
| HiPIP | High potential iron–sulfur protein |
| HOQNO | *n*-Heptylhydroxyquinoline *N*-oxide |
| $H_4FA$ | Tetrahydrofolic acid |
| $H_4MP$ | Tetrahydromethanopterin |
| MF | Methanofuran |
| MGD | Molybdopterin guanine dinucleotide |
| OTG | *n*-Octyl-β-D-thioglucoside |
| SOD | Superoxide dismutase |

# Chapter 1
# General Considerations

Organisms require energy for the life processes such as growth and movement, and the energy is supplied in the most cases by adenosine 5′-triphosphate (ATP). When ATP is hydrolyzed by the catalysis of ATPase to adenosine 5′-diphosphate (ADP), energy of 7.3 kcal is released per mole (ca. 500 g). The organisms utilize this energy for their life processes. In general, a man with a body weight of 60 kg needs about 1800 kcal per day in normal life without carrying out hard work or extensive movement. If the total daily energy required by him is supplied only by the hydrolysis of ATP to ADP, he needs 124 kg ATP [$1800 \div 7.3 = 247$ (moles); $0.5\ \text{kg} \times 247 = 124\ \text{kg}$]. It is impossible for him to hold a level of ATP of 124 kg in his body. The sum of the amounts of ATP and ADP present in one human body is only several tens of grams. Therefore, ADP formed by the hydrolysis of ATP has to be returned to ATP; ATP has to be used by a process of recycling. For this recycling, energy is necessary. To supply the energy, the human being needs to consume food, and bacteria also require nourishment. Furthermore, in the case of the photosynthetic organisms, light is required. However, in this book the photosynthetic organisms are not included except for a brief explanation. The food for bacteria varies according to the organism (Fig. 1.1).

The bacteria which feed organic compounds and oxidize them without utilization of light are called chemoheterotrophic bacteria. Although many chemoheterotrophic bacteria oxidize organic compounds with oxygen, some of them are known to use nitrate, sulfate, and ferric ions in place of molecular oxygen to oxidize organic compounds. Some bacteria oxidize inorganic compounds with molecular oxygen to acquire energy. These are called chemoautotrophic bacteria and, especially if they produce the cellular materials from carbon dioxide, chemolithoautotrophic bacteria. The inorganic compounds oxidized by chemolithoautotrophic bacteria are ammonia, nitrite, hydrogen sulfide (and other oxidizable sulfur compounds), ferrous ions, hydrogen, and others. Furthermore, some chemolithoautotrophic bacteria oxidize thiosulfate and ferrous ion with nitrate, and some oxidize hydrogen with carbon dioxide.

Some bacteria utilize light energy in similar fashion to plants and algae. These are called photosynthetic bacteria. Cyanobacteria oxidize water with light energy and produce molecular oxygen, like plants and algae; these are known as oxygenic

*Chemolithoautotrophic Bacteria*. T. Yamanaka
doi: 10.1007/978-4-431-78541-5_1, © Springer 2008

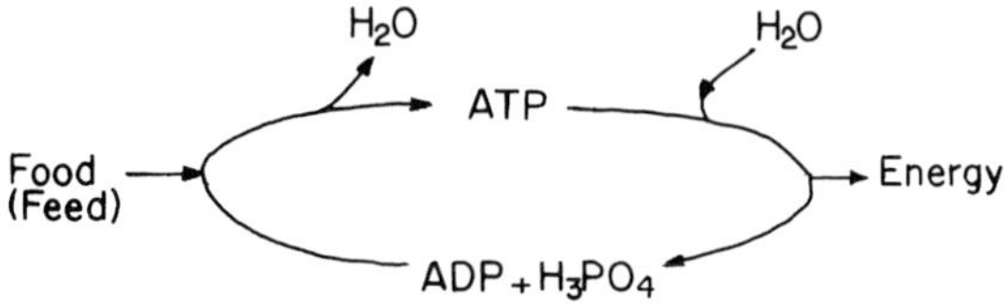


**Fig. 1.1.** A conceptual scheme showing how the organisms acquire the energy for the life processes

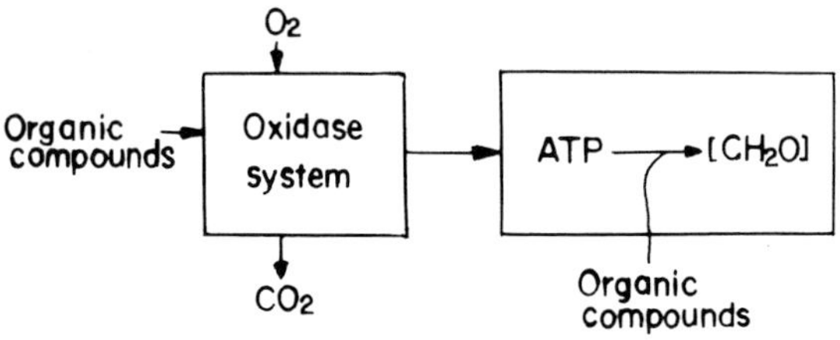


**Fig. 1.2.** A schematic presentation of the processes which yield energy and biosynthesize organic compounds in the chemoheterotrophic organisms. $[CH_2O]$ in this figure and in the following figures represents the organic compounds newly biosynthesized in the organisms

photosynthetic bacteria. Other photosynthetic bacteria do not produce molecular oxygen as the result of the photosynthesis and are called anoxygenic photosynthetic bacteria. Some oxidize inorganic compounds using light energy and are called photoautotrophic bacteria. Bacteria that oxidize organic compounds with light energy are called photoheterotrophic bacteria.

## 1.1 Chemoheterotrophic Bacteria

Many chemoheterotrophic bacteria oxidize organic compounds with molecular oxygen (Fig. 1.2). They oxidize many organic compounds such as carcasses and excreta of animals, and dead plants, and participate in the cleaning pollution from the surface of Earth.

Some chemoheterotrophic bacteria utilize inorganic compounds other than oxygen to oxidize organic compounds. The sulfate-reducing bacteria form ATP by oxidizing organic compounds with sulfate. In this case, sulfate is first changed to adenosine 5′-phosphosulfate (adenylyl sulfate, or APS) by an enzymatic reaction with ATP, and the resulting APS is used as the oxidant of the organic compounds. The bacteria of *Desulfovibrio* genus cannot oxidize acetic acid, while those of *Desulfotomaculum* genus oxidize the organic compounds to $CO_2 + H_2O$. The process by which the bacteria form ATP by oxidizing organic compounds with sulfate is called sulfate respiration (Fig. 1.3). The sulfate-reducing bacteria produce hydrogen sulfide as the result of the reduction of sulfate. This gas is toxic, and causes an environmental pollution because of its odor and corrosion of some metals. It is sometimes very dangerous at higher concentrations. Furthermore, when hydrogen sulfide thus formed is oxidized to sulfuric acid by the sulfur-oxidizing bacteria, the acid pollution and corrosion of concrete are caused in some cases.

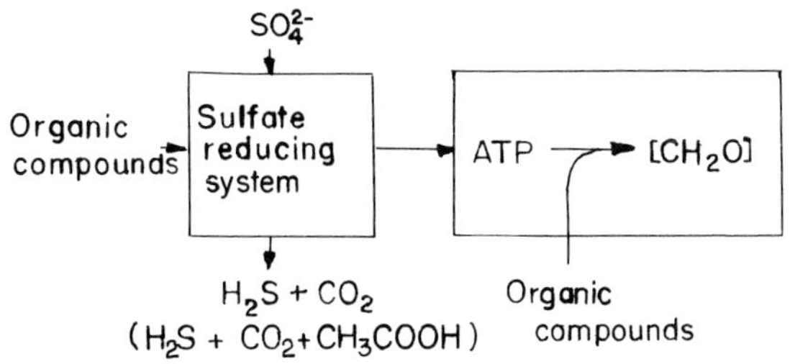


**Fig. 1.3.** A schematic presentation of the mechanisms which yield energy and biosynthesize organic compounds by the sulfate respiration

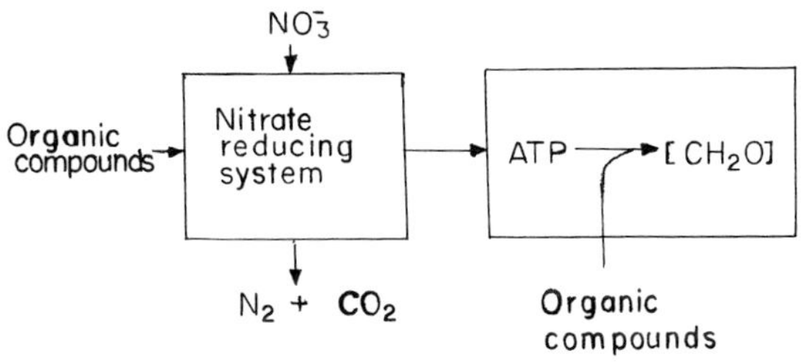


**Fig. 1.4.** A schematic presentation of the mechanisms which yield energy and biosynthesize organic compounds by the nitrate respiration

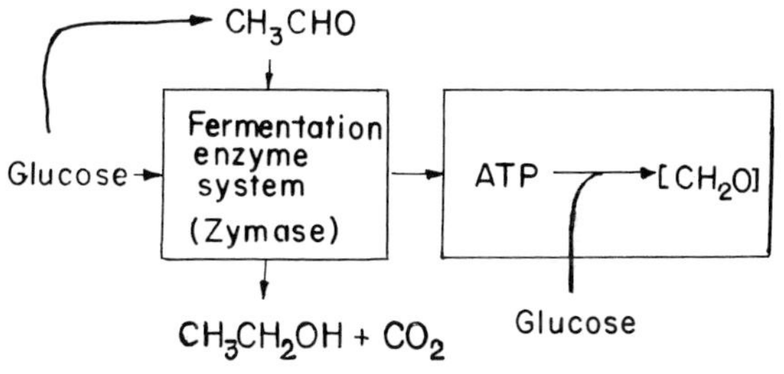


**Fig. 1.5.** A schematic presentation of the mechanisms which yield energy and biosynthesize organic compounds by the alcohol fermentation

The denitrifying bacteria oxidize organic compounds with nitrate and its reduced products to biosynthesize ATP (Fig. 1.4). Such processes are called nitrate respiration. In nitrate respiration, nitrate is reduced to nitrogen gas in many cases, through the process known as denitrification. Nitrate is a good nitrogen source for plants, while the denitrifying bacteria require it for biosynthesis of ATP. Therefore, the denitrifying bacteria and plants scramble for nitrate in some cases. In particular in the Torrid Zone, the denitrifying bacteria cause problems regarding the use of nitrogen fertilizer. The denitrifying bacteria reduce nitrate to nitrogen gas before plants can utilize it. The sulfate and nitrate respirers decompose anaerobically organic compounds and make the environment clean when oxygen is not available.

Fermentation is one of the chemoheterotrophic energy acquiring processes. In this process, organic compounds are anaerobically oxidized by organic compounds to produce ATP (Fig. 1.5). For example, in alcohol fermentation, glyceraldehyde-

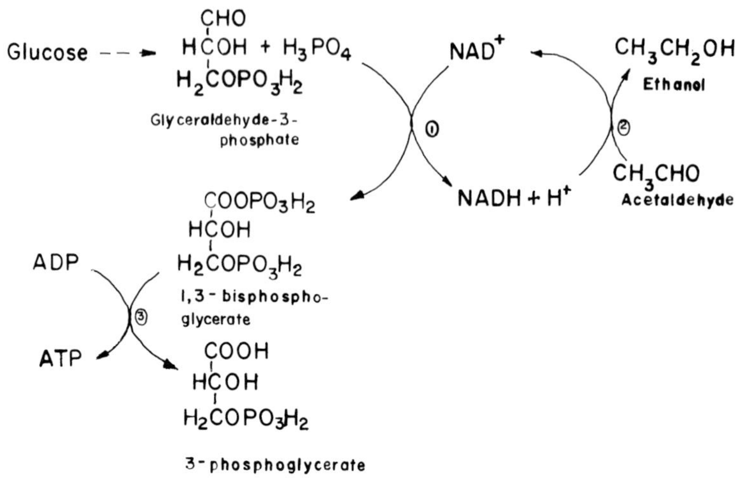


**Fig. 1.6.** An example of the reaction which yields energy by the oxidation of an organic compound by an organic compound in the alcohol fermentation. *Circled numbers*: *1*, glyceraldehyde-3-phosphate dehydrogenase (phosphorylating); *2*, alcohol dehydrogenase; *3*, phosphoglycerate kinase.

3-phosphate is oxidized by acetaldehyde, and as a result acetaldehyde is reduced to ethanol (Fig. 1.6). The fermentation also contributes to the decomposition of organic compounds and the cleansing of the environment when oxygen is not available.

## 1.2 Chemolithoautotrophic Bacteria

Autotrophic organisms grow only on inorganic compounds, the most well-known example of these being green plants. Considering plants, the organisms which can grow only on inorganic compounds may appear not to be special. However, the chemolithoautotrophic bacteria, unlike plants, do not utilize light energy and have to acquire all the energy for the life processes from the oxidation of inorganic compounds. Therefore, the mechanisms by which the chemolithoautotrophic bacteria acquire energy are different from those used by plants.

Organisms such as $C_1$-compund utilizing bacteria which grow on methanol, methylamine, etc., are sometimes included in the chemoautotrophic bacteria. The bacteria which can grow on only inorganic compounds and form their cellular materials from carbon dioxide are called chemolithoautotrophic bacteria. Some of the chemolithoautotrophic bacteria can grow not only autotrophically but also heterotrophically; these are known as facultative chemolithoautotrophic bacteria.

It was S.N. Winogradsky (1891) who discovered chemolithoautotrophic bacteria by verifying that the bacteria produce the cellular materials from carbon dioxide

(Brock and Schlegel, 1989). Namely, using ammonia oxidizing bacteria he found that the ratio of the amount of the inflammable material increased in the culture medium to the amount of ammonia oxidized was constant.

The chemolithoautotrophic bacteria participate intimately in the conversion and circulation of inorganic materials on Earth. In this book, mainly the properties of the chemolithoautotrophic bacteria and their relationship to the environment are described. The properties of the chemolithoautotrophic bacteria will be briefly surveyed below before their details are described in the corresponding chapters.

### *1.2.1 Ammonia-Oxidizing Bacteria*

These bacteria oxidize ammonia to nitrous acid via hydroxylamine (Fig. 1.7). A typical species of such bacteria is *Nitrosomonas europaea*. In addition to this, *Nitrosomonas eutrophus*, *Nitrosococcus oceanus*, *Nitrosospira briensis*, *Nitrosolobus multiformis* among others are known in this group.

In *N. europaea* ammonia ($NH_3$) is first oxidized by ammonia monooxygenase to produce hydroxylamine ($NH_2OH$) (Hofman and Lees, 1953), and molecular oxygen ($O_2$) is required in this reaction (Dua et al., 1979; Hollocher et al., 1981). Then, hydroxylamine formed is oxidized to nitrous acid by hydroxylamine oxidoreductase. In this reaction, oxygen atom of water is incorporated to hydroxylamine to produce nitrous acid ($HNO_2$) (Yamanaka and Sakano, 1980; Andersson and Hooper, 1983). However, molecular oxygen is usually necessary for the growth of the bacterium, because hydrogen atoms [H] (1.1) liberated from hydroxylamine and water should be oxidized to water.

$$NH_2OH + H_2O \rightarrow HNO_2 + 4[H] \tag{1.1}$$

### *1.2.2 Nitrite-Oxidizing Bacteria*

The bacteria grow by oxidizing nitrite ($NO_2^-$) to nitrate ($NO_3^-$) (Fig. 1.8). *Nitrobacter winogradskyi* is a typical species in this group. In addition to this, *Nitrobacter hamburgensis*, *Nitrospina gracilis*, *Nitrococcus mobilis* and others are known

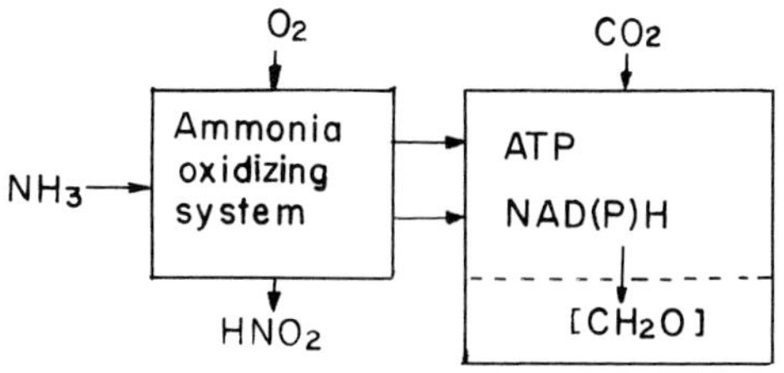


**Fig. 1.7.** A schematic presentation of the mechanisms by which the ammonia-oxidizing bacteria acquire energy and biosynthesize organic compounds

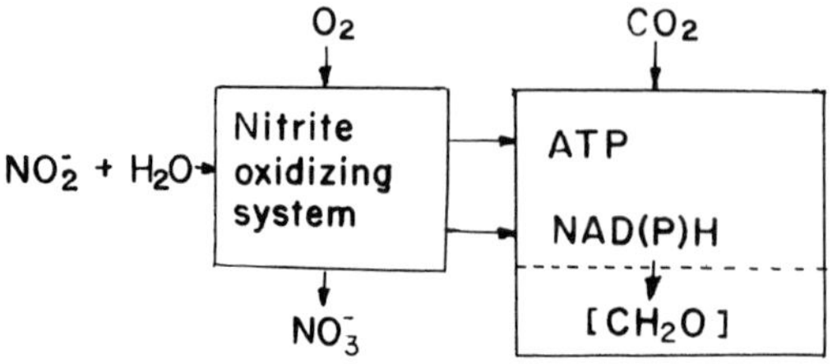


**Fig. 1.8.** A schematic presentation of the mechanisms by which the nitrite-oxidizing bacteria acquire energy and biosynthesize organic compounds

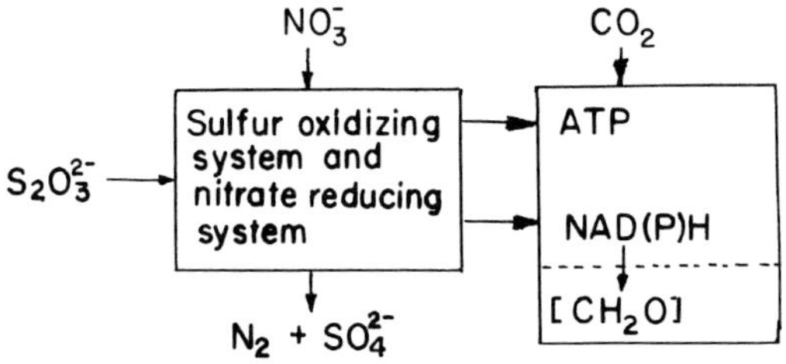


**Fig. 1.9.** A schematic presentation of the mechanisms by which *Thiobacillus denitrificans* acquires energy and biosynthesizes organic compounds

to belong this group. When *N. winogradskyi* oxidizes nitrite, oxygen atom of water is incorporated into nitrate formed (Aleem et al., 1965). However, molecular oxygen is necessary for the bacterium to oxidize hydrogen atoms [H] liberated from water.

$$NO_2^- + H_2O \rightarrow NO_3^- + 2[H] \quad (1.2)$$

The reaction (1.2) is catalyzed by nitrite oxidoreductase. This enzyme catalyzes actively the oxidation of nitrite at pH 8.0, while it catalyzes more preferentially the reduction of nitrate at pHs below 6.0. Therefore, the pH of the environments should be kept around 8 (Tanaka et al., 1983) for the bacterium to oxidize smoothly nitrite.

### 1.2.3 Denitrifying Bacteria

*Thiobacillus denitrificans* is well known as a lithoautotrophic denitrifier; it oxidizes anaerobically thiosulfate ($S_2O_3^{2-}$) with nitrate ($NO_3^-$) to biosynthesize ATP although it also aerobically oxidizes thiosulfate (Aminuddin and Nicholas, 1973, 1974a) (Fig. 1.9). The anaerobic oxidation system of thiosulfate is similar to those of other thiobacilli (though these are aerobic bacteria), and the denitrifying processes in the bacterium are similar to those in the heterotrophic denitrifiers. The complete genome of the bacterium has recently been sequenced (Beller et al., 2006).

Recently, other lithoautotrophic denitrifiers have been found that oxidize ferrous compounds with nitrate; *Ferroglobus placidus* (Hafenbradl et al., 1996) and *Dechlorosoma suillum* (Chaudhuri et al., 2001) (Fig. 1.10).

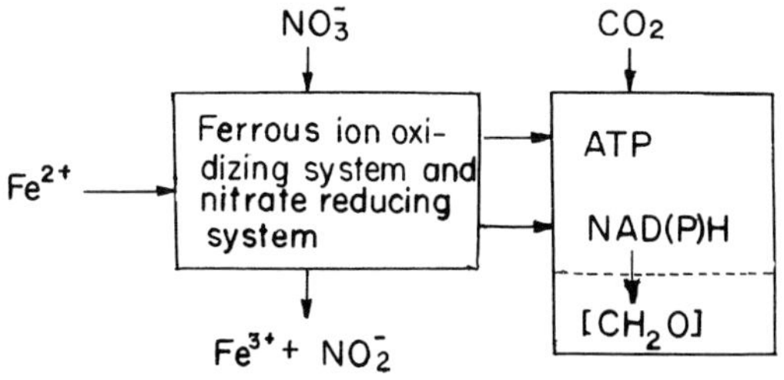


**Fig. 1.10.** A schematic presentation of the mechanisms by which *Ferroglobus placidus* and *Dechlorosoma suillum* acquire energy and biosynthesize organic compounds

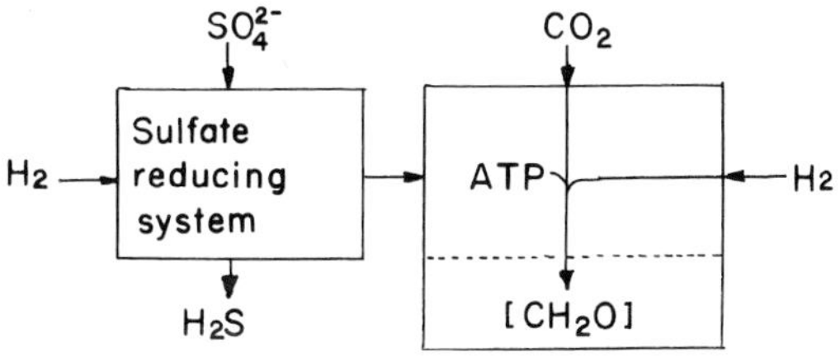


**Fig. 1.11.** A schematic presentation of the mechanisms which yield energy and biosynthesize organic compounds in the lithoautotrophic sulfate respiration. Note that organic compounds are oxidized with sulfate to yield energy, in the sulfate respiration shown in Fig. 1.3

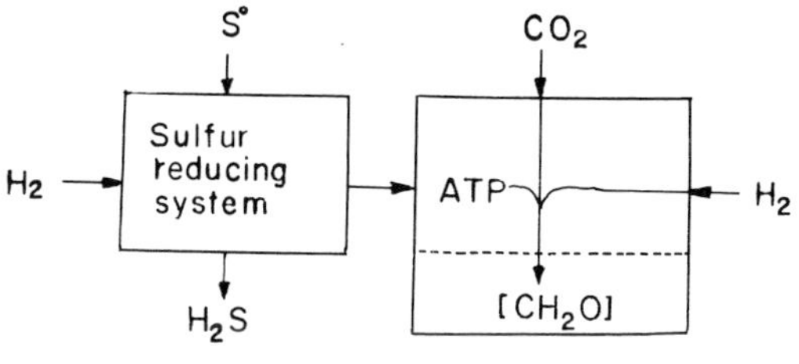


**Fig. 1.12.** A schematic presentation of the mechanisms which yield energy and biosynthesize organic compounds in the lithoautotrophic sulfur respiration

## *1.2.4 Sulfate-Reducing Bacteria*

Although most of the sulfate-reducing bacteria are heterotrophic, some lithotrophic sulfate-reducing bacteria are known; these bacteria grow by oxidizing hydrogen gas ($H_2$) with sulfate ($SO_4^{2-}$) (Fig. 1.11).

Recently, many hyperthermophiles have been found which grow by oxidizing hydrogen gas with elemental sulfur ($S^o$) (Stetter, 1994) (Fig. 1.12). The oxidation of hydrogen gas with elemental sulfur in the bacteria will probably proceed in a similar way to its oxidation with sulfate, although the details of the processes are not yet known. The process that biosynthesizes ATP by the oxidation of hydrogen gas with elemental sulfur is called sulfur respiration. The lithoautotrophic sulfate-reducing bacteria include *Desulfotomaculum geothermicum*, *Desulfobacterium autotrophicum*, *Desulfovibrio simplex*, and others (Fauque et al., 1991). Some strains of *Desulfovibrio vulgaris* also can grow autotrophically (Peck, 1960; Badziong and Thauer, 1978). *Pyrodictium occultum* and *Pyrobaculum islandicum* are

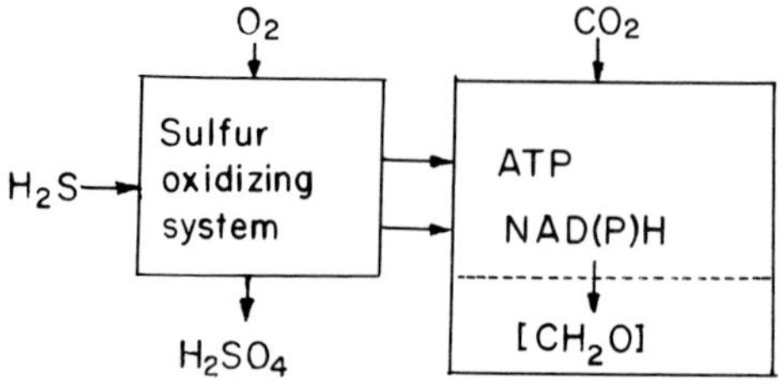


**Fig. 1.13.** A schematic presentation of the mechanisms by which the sulfur-oxidizing bacteria acquire energy and biosynthesize organic compounds

examples of the lithoautotrophic hyperthermophiles that oxidize hydrogen gas with elemental sulfur (Stetter, 1994).

### *1.2.5 Sulfur-Oxidizing Bacteria*

Sulfur-oxidizing bacteria grow by oxidizing mainly hydrogen sulfide ($H_2S$), elemental sulfur ($S^o$), thiosulfate ($S_2O_3^{2-}$) with molecular oxygen ($O_2$) [in some cases, with nitrate ($NO_3^-$) (cf. Fig. 1.9)] (Fig. 1.13).

There are two groups in the sulfur-oxidizing bacteria that are related to growth pH (Kuenen, 1989). The bacteria that grow within pH range from 1 to 5 include *Acidithiobacillus* (formerly *Thiobacillus*) *thiooxidans*, *Acidithiobacillus* (formerly *Thiobacillus*) *ferrooxidans*, *Acidithiobacillus* (formerly *Thiobacillus*) *acidophilus* and others, while the bacteria that grow in pH ranging from 5 to 8 include *Thiobacillus neapolitanus*, *Thiobacillus thioparus* and others. However, in reality *T. neapolitanus* can grow even at pH as low as 2.5. As *A. ferrooxidans* oxidizes also ferrous ion in addition to the sulfur compounds, it is known as the acidophilic iron oxidizing bacterium (see below). *Starkeya novella* (formerly *Thiobacillus novellus*), *Paracoccus* (formerly *Thiobacillus*) *versutus*, and *A. acidophilus* are facultative chemolithoautotrophic bacteria which grow not only on inorganic compounds but also on organic compounds. In particular, *P. versutus* is similar to the $C_1$-compound utilizing bacteria in having methylamine dehydrogenase (Vellieux et al., 1986). Some bacteria included in marine *Beggiatoa* genus grow lithoautotrophically by aerobic oxidation of hydrogen sulfide (Strohl, 1989). The lithoautotrophic species of *Beggiatoa* and *Thiobacillus* (or *Acidithiobacillus*) genera support the animal ecological system around the hydrothermal vents at the deep marine bottom (Childress et al., 1987; Jannasch et al., 1989).

### *1.2.6 Iron-Oxidizing and -Reducing Bacteria*

Among the acidophilic iron-oxidizing bacteria, *Acidithiobacillus ferrooxidans* (Kuenen, 1989) and *Leptospirillum ferrooxidans* (Eccleston et al., 1985) are well

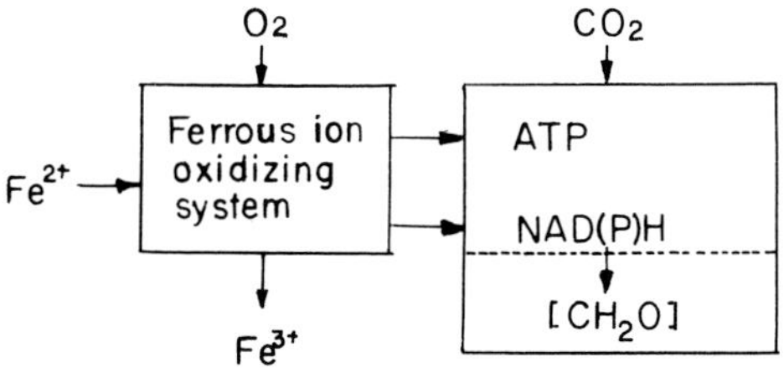


**Fig. 1.14.** A schematic presentation of the mechanisms by which the iron oxidizing bacteria acquire energy and biosynthesize organic compounds

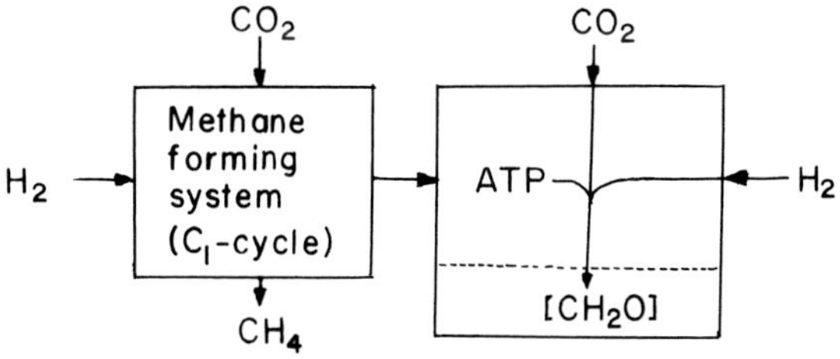


**Fig. 1.15.** A schematic presentation of the mechanisms by which the methanogens grow on hydrogen gas and carbon dioxide

known. They grow at pH 2.0 by oxidizing ferrous ion ($Fe^{2+}$) to ferric ion ($Fe^{3+}$) (Fig. 1.14). As already mentioned, *A. ferrooxidans* is also included in the sulfur-oxidizing bacteria.

*Gallionella ferruginea* grows by oxidizing ferrous ion at neutral pH (Hanert, 1989). As ferrous ion is easily oxidized spontaneously by molecular oxygen at pH 7.0, it appears very difficult for the bacterium to grow on ferrous ion at pH 7.0. Thus, when the bacterium is cultivated in the laboratory, the partial pressure of molecular oxygen should be limited to very low (1/100 of atmospheric concentration). However, the bacterium actually grows in nature, e.g., at the bottom of Crater Lake, Oregon, USA (Dymond et al., 1989). The bacteria grow there in a mat state.

Several photosynthetic bacteria have been discovered which oxidize ferrous ion anaerobically using light energy (Ehrenreich and Widdel, 1994; Heising et al., 1999). The investigation of these bacteria suggests that the banded iron oxide formation can occur even in the absence of molecular oxygen.

Bacteria included in *Geobacter* genus oxidize anaerobically organic compounds or inorganic compounds with ferric ion; as a result ferric ion is reduced to ferrous ion. By the actions of the bacteria, various ores of iron, e.g., magnetite, are produced (Chaudhuri et al., 2001).

## *1.2.7 Methanogens*

Methanogens (methane-producing bacteria) oxidize hydrogen gas ($H_2$) with carbon dioxide ($CO_2$) to produce methane ($CH_4$) (Fig. 1.15). Some methanogens utilize one carbon compound such as carbon monoxide (CO), methanol, methylamine, and formate in addition to carbon dioxide (Zeikus, 1977; Wolfe, 1980). Some methanogens utilize acetic acid to produce methane (Bhatnagar et al., 1991).

The methanogens which grow on hydrogen gas and carbon dioxide like *Methanobacterium thermoautotrophicum* ΔH are lithoautotrophs. Methanogens form methane to biosynthesize ATP; the content of cellular ATP is high when the bacteria actively produce methane (Blaut and Gottschalk, 1984). In the methanogens, the substrate level phosphorylation is not found during the formation of ATP, and the ATP synthesis is inhibited by ionophores (Blaut and Gottschalk, 1985). Therefore, the methane formation in the methanogens is not fermentation but is rather a carbon dioxide respiration (Wolfe, 1980; Daniels et al., 1984). The lithoautotrophic methanogens are unique in utilizing carbon dioxide both as the energy source and the source for the cellular materials.

Recently, *Pseudomonas* sp. HD-1 has been found, which grows on carbon dioxide and hydrogen gas and decomposes anaerobically *n*-alkane, benzene, and toluene (Morikawa and Imanaka, 1993).

# Chapter 2
# Cytochromes

When bacteria oxidize substrates with terminal electron acceptors such as molecular oxygen, sulfate, and nitrate, the electrons liberated from the substrates are transferred to the acceptors through cytochromes. Cytochromes belong to hemoproteins, but not all the hemoproteins are cytochromes. There are roughly two groups in the hemoproteins. Hemoglobin typifies one group while cytochrome typifies the other. Hemoglobin functions only when the iron in the heme is ferrous state [Fe(II)], while cytochrome functions by varying the iron valency in the heme. The heme iron of cytochrome varies in most cases between ferrous iron [Fe(II)] and ferric iron [Fe(III)]; cytochrome functions by oxidation and reduction of the heme iron.

## 2.1 Hemes

As there are many kinds of hemes which are different in their structures (Fig. 2.1), there are also many kinds in cytochromes which have hemes as the prosthetic group. Heme is a complex of porphyrin with iron. In Fig. 2.1, the structures of several hemes are shown which will appear in this book.

## 2.2 Kinds of Cytochromes

Cytochromes are classified according to the heme which they have as the prosthetic groups. In Table 2.1, the relationships between main cytochromes and hemes are shown.

### *2.2.1 Heme A-Containing Cytochromes*

Cytochromes are classified according to the hemes as the prosthetic group. Most cytochromes having heme A show activity to reduce molecular oxygen except for

*Chemolithoautotrophic Bacteria*. T. Yamanaka
doi: 10.1007/978-4-431-78541-5_2, © Springer 2008

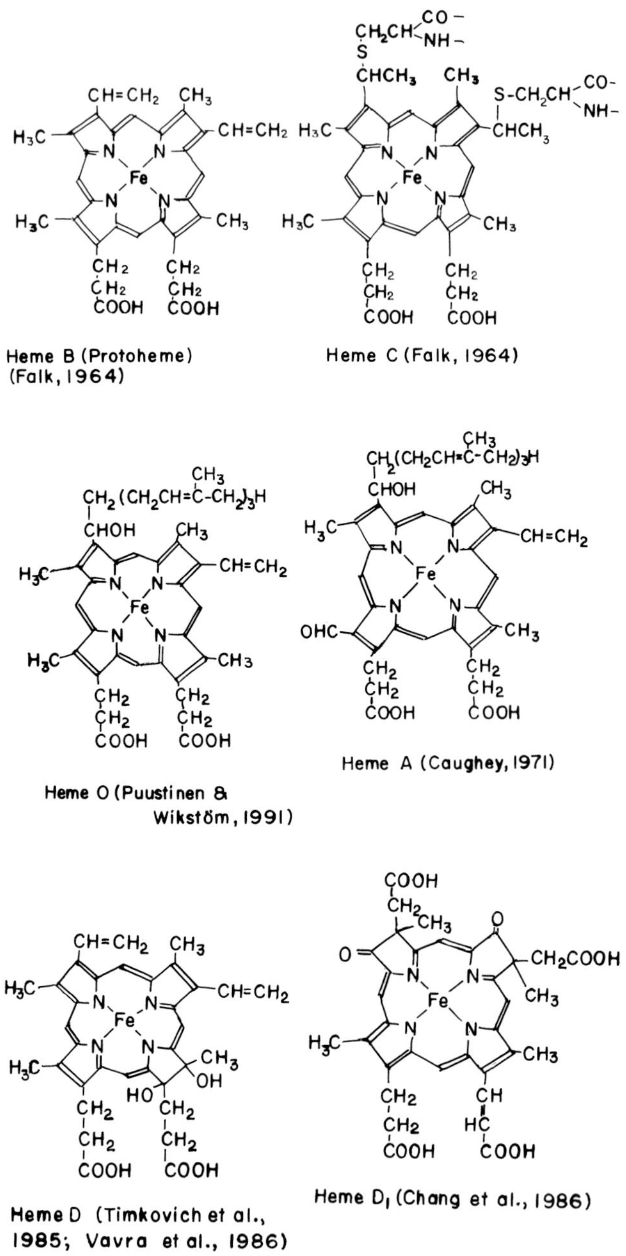


**Fig. 2.1.** Structures of various hemes found in bacteria

**Table 2.1.** Brief properties of several cytochromes obtained from bacteria and their relations to hemes (prepared based on Yamanaka, 1992; Fukaya et al., 1993; Richter et al., 1994; Tamegai and Fukumori, 1994; Fujiwara and Fukumori, 1996; Berben, 1996; Igarashi et al., 1997; Qureshi et al., 1998)

| Heme | Cytochrome | Molecular mass of monomer (kDa) | Characteristics |
|---|---|---|---|
| A | $aa_3$ | 67 (86.5)[a] (*N. winogradskyi*) | Cytochrome *c* oxidase without proton pumping activity. Contains 2 heme A molecules and 2 Cu atoms |
| | $a_3$ | 55 (*S. novella*) | Cytochrome *c* oxidase with proton pumping activity (dimer). Contains 1 heme A molecule and 1 Cu atom |
| B | $b$ | 50 | Electron transfer component. Contains 2 heme B molecules and occurs as cytochrome $bc_1$ |
| C | $c$ | 12.4 (*N. winogradskyi*) | Electron transfer component. Contains 1 heme C molecule |
| | $c$ | 9.1 (*N. europaea*) | Electron transfer component. Contains 1 heme C molecule |
| | $c_3$ | 12.3 (*D. vulgaris*) | Electron transfer component. Contains 4 heme C molecules |
| | $NH_2OH$ oxidoreductase | 190 | Catalyzes oxidation of $NH_2OH$ to $HNO_2$. Contains 21 heme C and 3 heme P-460 molecules |
| In addition to these, cytochromes $c_1$, $c_2$, $c_4$, $c_5$, $c_6$, $c_8$, and *f* are known all of which function as the electron transfer components. | | | |
| A + B | $ba_3$ | 53 (*T. thermophilus*) | Cytochrome *c* oxidase. Contains 1 heme A molecule, 1 heme B molecule and 2 Cu atoms |
| | $ba_3$ (or $ba_1$) | 140 (*Acetobacter aceti*) | Quinol oxidase |
| | $ba_3$ | 143 (*P. denitrificans*) | Quinol oxidase |
| A + C | $a_1c_1$ | 250 (*N. winogradskyi*) | Nitrite oxidoreductase |
| B + C | *cbb*[a] | 49 (*P. denitrificans*) | NO reductase |
| | $cbb_3$ | 105 (*M. magnetotacticum*) | Cytochrome *c* oxidase |
| | *ccb* | 197.5 (*Shewanella benthica*) | Quinol oxidase |
| B + O | $bo_3$ | 215 (*Escherichia coli*) | Quinol oxidase with proton pumping activity |
| B + D | *bd* | 100 (*E. coli*) | Quinol oxidase without proton pumping activity |
| C + $D_1$ | $cd_1$ | 60[b] (*Pseudomonas aeruginosa*) | Nitrite reductase |

In addition to the hemes described in Fig. 2.1 and this table, siroheme of sulfite reductase will be described in Chapter 4 (p. 59).

[a] NO reductase also occurs which lacks heme C (Suharti et al., 2001; de Vries et al., 2003)

[b] Deduced from DNA sequence

nitrite oxidoreductase (see e.g. Yamanaka, 1992). Cytochrome $aa_3$ has 2 molecules of heme A in one molecule. The cytochrome functions as cytochrome *c* oxidase which catalyzes the reduction of molecular oxygen with ferrocytochrome *c* and the translocation of proton ($H^+$) through the molecule, i.e., as a result, a gradient in proton concentration occurs through the mitochondrial inner membranes and the bacterial plasma membranes (proton pumping activity). The proton pumping activity is directly responsible for the biosynthesis of ATP. The cytochrome is distributed among organisms including human, animals, plants, and many aerobic microorganisms.

One of 2 molecules of heme A in the cytochrome $aa_3$ molecule reacts with molecular oxygen, carbon monoxide, and other ligands, but the other heme molecule does not react with these ligands. The protein moiety having the former heme is called cytochrome $a_3$ (component), while that having the latter heme is called cytochrome *a* (component). In addition to the two heme molecules, the cytochrome has 2–3 copper atoms. One copper atom which is functionally related to the cytochrome $a_3$ component and forms $a_3$-CuB site or dimetallic center is called CuB, while the other one or two copper atoms are called CuA. The cytochrome reduces molecular oxygen to water at the $a_3$-CuB site with the electrons transferred from ferrocytochrome *c* through CuA and cytochrome *a*. This electron pathway is found in cytochromes $aa_3$ of the mitochondrion and *Paracoccus denitrificans* (Tsukihara et al., 1995, 1996; Iwata et al., 1995). However, some bacterial cytochromes $aa_3$ are known which do not have CuA and still catalyze the reduction of molecular oxygen with ferrocytochrome *c*. Therefore, it has not yet been decided whether or not CuA is really necessary for cytochrome $aa_3$ to reduce molecular oxygen with ferrocytochrome *c* (Numata et al., 1989; Garcia-Horsman et al., 1994). On the contrary, it is known that cytochrome $aa_3$ can have no CuA and catalyze the reduction of molecular oxygen with quinol (quinol oxidase) but not with ferrocytochrome *c* (Lauraeus et al., 1991).

In addition to cytochrome $aa_3$, cytochromes $ba_3$, $baa_3$, and *aco* (or $cao_3$) are known, which have heme A together with other types of heme, and function as cytochrome *c* oxidases (see Yamanaka, 1992). Furthermore, cytochrome $a_1c_1$ occurs, which has two molecules of each of hemes A and C, one molybdenum atom, and 5 $[Fe_4S_4]$ clusters in the molecule. The cytochrome functions as nitrite oxidoreductase; it catalyzes the oxidation of nitrite to nitrate (Tanaka et al., 1983; Fukuoka et al., 1987; Suzuki et al., 1997).

### 2.2.2 Heme B-Containing Cytochromes

Cytochrome with heme B or protoheme as the prosthetic group is known as cytochrome *b*. Cytochrome *b* functions usually as the electron transfer component. However, the cytochrome system which participates in the oxidation of inorganic compounds in the chemolithoautotrophic bacteria is usually composed of cytochrome *c* and the terminal oxidase without cytochrome *b* (Yamanaka, 1996).

Among the bacterial terminal oxidases, there are some which have heme B such as cytochromes *ba*$_3$ (Zimmermann et al., 1988; Matsushita et al., 1990; Fukaya et al., 1993; Richter et al., 1994), *baa*$_3$ (Fujiwara et al., 1992) and *cbb*$_3$ (Garcia-Horsman et al., 1994; Tamegai and Fukumori, 1994). Cytochromes *baa*$_3$ and *cbb*$_3$ are cytochrome *c* oxidases, while some cytochromes *ba*$_3$ function as cytochrome *c* oxidase and some function as quinol oxidase. Cytochromes *ba*$_3$ and *baa*$_3$ are structurally similar to cytochrome *aa*$_3$, while cytochrome *cbb*$_3$ does not have CuA but still functions as a cytochrome *c* oxidase (Gray et al., 1994; Garcia-Horsman et al., 1994).

## 2.2.3 Heme C-Containing Cytochromes

The cytochromes with heme C as the prosthetic group are classified as cytochromes *c* (see e.g. Pettigrew and Moore, 1987; Moore and Pettigrew, 1990; Yamanaka, 1992). The cytochrome group includes cytochromes *c*, $c_2$, $c_3$, $c_5$, $c_6$, $c_8$, *f* and others, and they are all electron transfer components. Cytochrome *c* is widely distributed both in eukaryotes and many micro-organisms. Cytochrome $c_2$ occurs in the non-sulfur photosynthetic bacteria, while cytochrome $c_3$ occurs in the sulfate reducing bacteria, and this cytochrome contains four heme C molecules in one molecule. Cytochromes $c_6$ and $c_8$ are similar to each other in molecular mass of ca. 9 kDa, but they differ in the amino acid sequence of certain portions (Ambler, 1991). However, they resemble each other in their reactivity with several redox enzymes (Yamanaka, 1975). Cytochromes $c_4$ and $c_5$ were first found in *Azotobacter vinelandii*, and are now found to occur in several other bacteria. Cytochrome $c_4$ is a monomeric diheme cytochrome *c* with molecular mass of 20 kDa, while cytochrome $c_5$ is a dimer of a monoheme cytochrome *c* of 10 kDa (Meyer and Kamen, 1982). Hydroxylamine oxidoreductase from the ammonia oxidizing bacteria has 21 heme C molecules and 3 molecules of a variant heme C (heme P-460) in the molecule (Arciero et al., 1993; Igarashi et al., 1997).

## 2.2.4 Heme $D_1$-Containing Cytochromes

No cytochrome has been found that has only heme $D_1$ in the molecule. Cytochrome having hemes $D_1$ and C in one molecule is known: cytochrome *cd*$_1$ (Yamanaka and Okunuki, 1963a,b,c). Cytochrome *cd*$_1$ is the first cytochrome found to have different hemes in the molecule (Yamanaka, 1992). The molecular mass of this cytochrome is about 60 kDa. Although the cytochrome molecule is composed of one polypeptide, there are two domains in the molecule: heme $D_1$ binding domain and heme C binding domain (Fulop et al., 1995). This cytochrome is a nitrite reductase (Yamanaka, 1964) which is present in many denitrifying bacteria.

### 2.2.5 Heme O-Containing Cytochromes

Cytochrome having hemes O and B is known as cytochrome *bo*$_3$ (or *bo*). This cytochrome functions as the terminal oxidase in many bacteria. It resembles cytochrome *aa*$_3$ in structure; heme O and Cu$_B$ of the cytochrome compose the dimetallic center to reduce molecular oxygen (Mogi et al., 1998). However, this cytochrome does not utilize ferrocytochrome *c* as the electron donor. The electron donor for cytochrome *bo*$_3$ is ubiquinol; it is an ubiquinol oxidase. Moreover, it shows proton pumping activity.

In addition to the ubiquinol oxidase mentioned above, cytochrome *c* oxidase is known, which contains heme O, as already described. Cytochrome *aco* (or *cao*$_3$) has one molecule each of hemes A, B and O in the molecule, and shows cytochrome *c* oxidase activity (Qureshi et al., 1990). cytochromes *co* purified from *Pseudomonas aeruginosa* (Matsushita et al., 1982), *Methylophilus methylotrophus* (Froud and Anthony, 1984) and *Pseudomonas stutzeri* (Heiss et al., 1989) also act as cytochrome *c* oxidase.

### 2.2.6 Heme D-Containing Cytochromes

A cytochrome having hemes D and B in the molecule is known: cytochrome *bd*. This cytochrome was first purified from *Photobacterium phosphoreum* (Watanabe et al., 1979). It is a quinol oxidase and functions as the terminal oxidase in some bacteria. It utilizes menaquinol in vivo as the electron donor, while it catalyzes the oxidation of both menaquinol and ubiquinol in vitro, but it does not catalyze the oxidation of ferrocytochrome *c*. It does not have copper, but still reduces molecular oxygen at the site of heme D (Mogi et al., 1998).

# Chapter 3
# Nitrogen Circulation on Earth and Bacteria

About 78% of atmospheric gas of Earth is nitrogen ($N_2$). Many readers of this book may think that nitrogen gas in the atmosphere is unchangeable forever as it is a stable substance. However, when they know that proteins of animals, plants, and bacteria contain nitrogen atoms, they will understand that nitrogen atoms are present also in the organisms in other forms than gas. Ammonia ($NH_3$) is formed when dead bodies of animals (carcasses), their excreta, and dead plants are decomposed by bacteria. Ammonia thus formed and released from nitrogen fertilizer (ammonium sulfate, urea, etc.) is partially absorbed by plants, and its remaining part which has not been utilized by plants is oxidized to nitrite ($NO_2^-$) by ammonia-oxidizing bacteria. Resulting nitrite is not utilized by plants because it is poisonous. Nitrite is quickly oxidized to nitrate ($NO_3^-$) by nitrite-oxidizing bacteria. Through these processes, ammonia is actually oxidized to nitric acid, and nitric acid formed reacts with calcium carbonate in soil to form calcium nitrate. Nitrate is a good nitrogen source for all plants. Nitrate remained unabsorbed by plants is reduced to nitrogen gas by denitrifying bacteria. In some cases, the denitrifying bacteria compete with plants in utilizing nitrate.

If all ammonia and nitrate on Earth are reduced to nitrogen gas, plants on Earth perish, because plants cannot utilize nitrogen gas as the nitrogen source. However, it is fortunate for us that bacteria occur which reduce nitrogen gas to ammonia: nitrogen-fixing bacteria. In the bacteria, ammonia formed by the reduction of nitrogen gas instantly combines with organic compounds to form amino acids. In any case, nitrogen circulates as ammonia → nitrite → nitrate → nitrogen gas → ammonia (Fig. 3.1), and special bacteria participate in each reaction step.

In Fig. 3.1, industrial nitrogen fixation means industrial production of ammonium sulfate from nitrogen gas. Artificial oxidation of nitrogen means the formation of nitrogen oxides included in car exhaust and in smoke discharged from industrial chimneys. Furthermore, nitrogen oxides are produced by thunder. The nitrogen oxides thus formed are dissolved in rain water, become nitrous and nitric acids, and fall on Earth. The acidic rain thus formed is first a good nitrogen source for plants, and plants at first grow prosperously, but meanwhile soil becomes acidic and the plants die. Accumulated nitrate, as seen in Fig. 3.1, means nitrate adhered on the surface of walls of stone buildings observed mainly in Europe. The nitrate is located

*Chemolithoautotrophic Bacteria*. T. Yamanaka
doi: 10.1007/978-4-431-78541-5_3, © Springer 2008

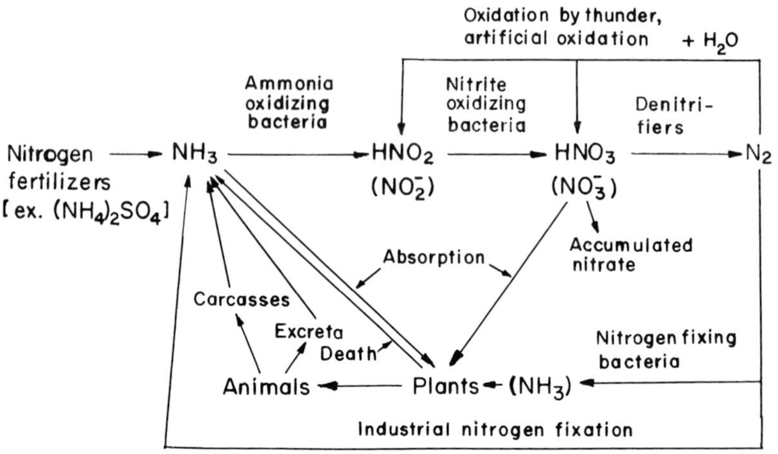


**Fig. 3.1.** A summary of natural circulation of nitrogen

temporarily out of the circulation of nitrogen. The natural niter and soda niter [naturally occurring potassium nitrate ($KNO_3$) and sodium nitrate ($NaNO_3$), respectively] are also states of nitrogen which are temporarily out of circulation. In the Edo era in Japan, niter or potassium nitrate was produced by using nitrifying bacteria. In this case, the nitrogen was also very temporarily out of circulation. As the niter was used as an ingredient of gun powder, it was meanwhile decomposed to generate nitrogen gas.

## 3.1 Bacterial Nitrification

### *3.1.1 Oxidation of Ammonia*

Some plants utilize ammonia or ammonium ion as a good nitrogen source. In particular, rice plants preferentially utilize ammonia as the nitrogen source when they are young, i.e., before flowering. However, ammonia is not necessarily utilized by all plants, while nitrate is utilized by any plant as the nitrogen source. Ammonia which has not been utilized by plants is first oxidized to nitrous acid ($HNO_2$) by the ammonia-oxidizing bacteria. As nitrous acid is usually changed to calcium nitrite by the reaction with calcium carbonate in soil, soil does not become acidic in most cases. However, if nitrous acid is formed so vigorously that the calcium carbonate cannot neutralize it, pollution due to nitrous acid can occur as described below.

Several ammonia-oxidizing bacteria are known, including *Nitrosomonas europaea*, *Nitrosomonas eutropha*, *Nitrosococcus oceanus*, *Nitrosospira briensis*,

*Nitrosovibrio tenuis*, and *Nitrosolobus multiformis* (Watson et al., 1989; Koops et al., 1991). Among these, *N. europaea* has been the most studied.

The ammonia-oxidizing bacteria oxidize ammonia to nitrous acid via hydroxylamine ($NH_2OH$) (Lees, 1952; Hofman and Lees, 1953); ammonia is first oxidized to hydroxylamine by the catalysis of ammonia monooxygenase (AMO) (Dua et al., 1979; Hollocher et al., 1981). In this reaction, molecular oxygen is utilized. Then, hydroxylamine formed is oxidized to nitrous acid by the catalysis of hydroxylamine oxidoreductase (HAO).

$$NH_3 + O_2 + 2[H] \xrightarrow{AMO} NH_2OH + H_2O \tag{3.1}$$

$$NH_2OH + H_2O \xrightarrow{HAO} HNO_2 + 4H^+ + 4e \tag{3.2}$$

[[H], hydrogen atom bound to some compound which has not yet been known (Suzuki & Kwok, 1981); *e*, electron, and should be finally oxidized by molecular oxygen]

Ammonia monooxygenase is very unstable; when the cells of *N. europaea* are destroyed the enzyme is mostly inactivated (Suzuki and Kwok, 1970; Suzuki et al., 1981). Therefore, little is known about the properties of the enzyme. The copper atom seems important for the enzyme to function, as the enzyme is inactivated by many cuprous chelating agents (Hooper and Terry, 1973; Wood, 1986), and the enzyme seems to be activated by cuprous ions (Ensign et al., 1993). Moreover, the enzyme purified from the heterotrophic nitrifier *Paracoccus denitrificans* has been found to be activated by the cuprous ion (Moir et al., 1996). However, the *P. denitrificans* enzyme may be a little different from the *N. europaea* enzyme, because it is not inhibited by acetylene (Moir et al., 1996), while the *N. europaea* enzyme is inhibited by the compound (Hooper and Terry, 1973). Electron paramagnetic resonance (EPR) studies show that iron is also important for the function of the enzyme (Zahn et al., 1996). The amino acid sequence of the *N. europaea* enzyme is deduced from DNA (Hyman and Wood, 1985; McTavish et al., 1993; Bergmann and Hooper, 1994), although firm evidence has not been obtained that the DNA really encodes the enzyme.

Ammonia monooxygenase requires 2[H] (two hydrogen atoms or electrons) to oxygenate ammonia, as shown by equation (3.1). As the oxidation of ammonia to hydroxylamine by the cell-free extracts of *N. europaea* is activated by addition of cytochrome *c*-554 (Yamanaka and Shinra, 1974), the 2 [H] are thought to be supplied to the enzyme through the cytochrome (Suzuki and Kwok, 1981). Incidentally, the substrate for ammonia monooxygenase is ammonia, but not ammonium ion (Suzuki et al., 1974).

Ammonia monooxygenase catalyzes also the oxidation of methane (Hyman and Wood, 1983) and carbon monoxide (Tsang and Suzuki, 1982) to methanol and carbon dioxide, respectively, in addition to the catalysis of the oxidation of ammonia. The oxidation of methane by the enzyme will be described again below in relation to the regulation of the methane formation by the ammonia-oxidizing bacteria.

**Table 3.1.** Inhibitors to ammonia oxidizing activity of *Nitrosomonas europaea* (Prepared based on: Hooper and Terry, 1973; Powell and Prosser, 1985; Bedard and Knowles, 1989; Juliette et al., 1993; Roy and Knowles, 1995; Matsuba et al., 2003)

| Inhibitors | Minimal concentration to cause more than 70% inhibition (μM) |
|---|---|
| Sodium diethyldithiocarbamate [$(CH_2CH_3)_2NCS_2Na$] | 10 |
| Thiourea [$H_2NCSNH_2$] | 37.4 |
| Allylthiourea [$CH_2{=}CHCH_2NHCSNH_2$] | 1.0 |
| Nitrapyrin [2-chloro-6-(trichloromethyl)pyridine] 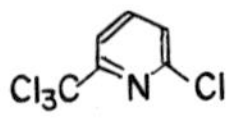 | 1.0 |
| Br-MAST (2-amino-4-tribromomethyl-6-trichloromethyl-1,3,5-triazine) 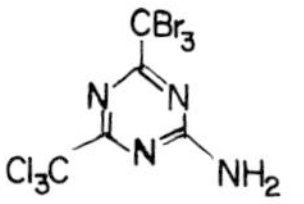 | 0.076 (50% inhibition) |
| Chloropicolinic acid 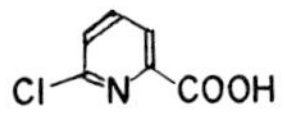 | 2.2 (60% inhibition) |
| Allylsulfide [$(CH_2{=}CHCH_2)_2S$] | 0.4 (complete inhibition) |

To clean up the pollution caused by ammonia, this compound should be changed finally to nitrogen gas by the actions of the ammonia-oxidizing bacteria and other two kinds of bacteria described below. Meanwhile, in agriculture, ammonia supplied as the nitrogen fertilizer should be utilized by plants (crops) before being reduced to nitrogen gas. For this purpose, the oxidation of ammonia by the ammonia-oxidizing bacteria should be restrained by addition to the fertilizer of nitrification-controlling reagents such as thiourea, allylthiourea, nitrapyrin, and others (Hooper and Terry, 1973; Powell and Prosser, 1985) (Table 3.1). These compounds are inhibitors of ammonia monooxygenase. Acetylene is also an inhibitor of the enzyme, but its application to crops is difficult as it is an inflammable gas. Although the use of the nitrification-controlling reagents restrains the bacterial oxidation of ammonia and saves nitrogen fertilizer, it supports the pollution caused by ammonia. In short, the bacterial cleaning of the pollution caused by ammonia conflicts with the use of the nitrogen fertilizer in agriculture.

## *3.1.2 Oxidation of Hydroxylamine*

### (a) Hydroxylamine Oxidoreductase

The oxidation of hydroxylamine by the ammonia-oxidizing bacteria is catalyzed by hydroxylamine oxidoreductase (Yamanaka and Sakano, 1980). The molecular

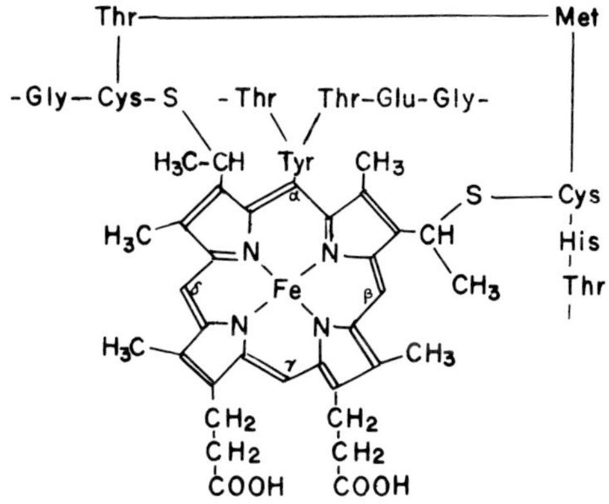


**Fig. 3.2.** Structure of heme P-460 (Prepared on the basis of Arciero et al., 1993; Sayavedra-Soto et al., 1994; Igarashi et al., 1997). *Cys* means $-CH_2CH(NH^-)CO^-$; *Gly*, *Glu*, *His*, *Met*, *Thr*, and *Tyr* are residues of glycine, glutamic acid, histidine, methionine, threonine, and tyrosine, respectively

mass of the enzyme is about 190 kDa, and its molecule contains 21 molecules of heme C and 3 molecules of heme P-460 (Arciero et al., 1993; Sayavedra-Soto et al., 1994). The three-dimensional structure of the enzyme has been elucidated by Igarashi et al. (1997). The enzyme molecule is composed of three subunits each of which contains 7 heme C molecules and 1 heme P-460 molecule.

Heme P-460 is a modified heme C in which a tyrosyl residue is bound to the α-position of heme C (Fig. 3.2). The tyrosyl residue bound to heme P-460 comes from the polypeptide composing the next subunit, so that the heme connects two adjacent subunits. This explains that the three subunits of the enzyme are separated from each other only after the enzyme has been treated with the reagents to destroy thioether linkage. The heme irons of all the heme C molecules in the enzyme have two histidine residues as the axial ligands. However, the redox potentials of the hemes vary widely according to the hemes (Prince and Hooper, 1987; Collins et al., 1993), and most of the heme molecules show the α absorption peak at 553 nm while some show the peak at 559 nm (Yamanaka et al., 1979b; Collins et al., 1993). These may be attributable to the differences in the protein environments around the hemes. The axial ligand to the heme iron of heme P-460 at the fifth position is histidine residue, while the sixth position of the heme iron is empty and hydroxylamine will bind to this position (Igarashi et al., 1997).

Besides heme P-460 of hydroxylamine oxidoreductase, cytochrome P-460 has been isolated from *N. europaea* (Erickson and Hooper, 1972; Miller et al., 1984; Numata et al., 1990). Cytochrome P-460 shows a weak activity of hydroxylamine oxidoreductase, and its molecule is composed of 3 subunits of 18 kDa. As the amino acid sequence of the cytochrome differs from that of hydroxylamine oxidoreductase and DNA encoding the cytochrome is found to be different from that of the oxidoreductase, the cytochrome is not a proteolytic fragment of the oxidoreductase (Bergman and Hooper, 1994b). Indeed, cytochrome P-460 has recently been crystallized and its spatial structure has been determined (Pearson et al., 2007). Its heme is a modified heme C in which lysine residue links to γ-carbon of the heme, and the heme shows the spectral properties analogous to those of heme P-460 in hydroxylamine oxidoreductase. Cytochrome P-460, which has the activity to catalyze the oxidation of hydroxylamine, has been obtained from the methane-oxidizing bacterium, *Methylococcus capsulatus* Bath (Zahn et al., 1994).

A question occurs as to why the bacterial enzyme has such a complicated structure, because hydroxylamine is oxidized to nitrite by the catalysis of ferric ion under aerobic conditions. In the nonenzymatic reaction, molecular oxygen is incorporated into nitrite formed by the oxidation of hydroxylamine, while the oxygen atom of water is incorporated into nitrite formed by the enzymatic oxidation of hydroxylamine (see below) (Yamanaka and Sakano, 1980; Andersson and Hooper, 1983). The mechanism in the bacterial oxidation of hydroxylamine will have been devised to reserve efficiently the energy of the reaction for the biosynthesis of adenosine triphosphate (ATP).

Hydroxylamine oxidoreductase was first purified by Hooper and Nason in 1965. They found that the enzyme catalyzed the reduction of horse ferricytochrome *c* with hydroxylamine but they found little nitrous acid formed as the result of the reaction; nitrite formed was approximately 5% as much as cytochrome *c* reduced in 0.1 M glycine–NaOH buffer, pH 9.8. As the ammonia-oxidizing bacteria oxidize hydroxylamine to nitrite, they thought that one or more enzymes in addition to the oxidoreductase might participate in the oxidation of hydroxylamine to nitrite or additional factor(s) might be necessary for changing hydroxylamine to nitrite (Hooper et al., 1977).

In 1980, a student for whom the author was a supervisor at Osaka University found that if the amount of ferricytochrome *c* used was 10 times as much as that of hydroxylamine added and the reactions were performed in 0.1 M phosphate buffer, pH 8.0, the enzyme catalyzed almost completely the oxidation of hydroxylamine to nitrite (Yamanaka and Sakano, 1980). Thus, it has been established that hydroxylamine is oxidized to nitrite by the catalysis of hydroxylamine oxidoreductase itself. Furthermore, the formation of nitrite by the enzymatic oxidation of hydroxylamine has been found to occur even under anaerobic conditions if a sufficient amount of the electron acceptor is present. Therefore, the enzymatic oxidation of hydroxylamine itself does not require molecular oxygen, though ferrocytochrome *c*-554 [native ferrocytochrome *c*] formed by the dehydrogenation of hydroxylamine has to be eventually oxidized by atmospheric oxygen, in vivo.

Why is so much ferricytochrome *c* necessary and molecular oxygen unnecessary for the enzymatic oxidation of hydroxylamine to nitrite? The author believes that the enzymatic oxidation of hydroxylamine occurs in two steps (see the reaction formulas below). At the first step (3.3), hydroxylamine is oxidized to an intermediate metabolite "NOH" by losing two hydrogen atoms, and at the second step (3.4), two hydrogen atoms are taken off from "NOH" and water. For the reactions (3.3) and (3.4), molecular oxygen is unnecessary. Assuming that the reaction (3.3) proceeds much more rapidly than the reaction (3.4), ferricytochrome *c* is almost exhausted at the reaction (3.3) if the amount of ferricytochrome *c* added is less than twice that of hydroxylamine added, and little electron acceptor is available for the reaction (3.4). Therefore, when the amount of the electron acceptor added is less than four times that of hydroxylamine, this compound is not oxidized completely to nitrite although, actually, a small amount of nitrite is formed because the reaction (3.4) will occur to some extent.

$$NH_2OH + 2Cyt\ c(Fe^{3+}) \xrightarrow{HAO} \text{“NOH”} + 2Cyt\ c(Fe^{2+}) + 2H^+ \qquad (3.3)$$

$$\text{“NOH”} + H_2O + 2Cyt\ c(Fe^{3+}) \xrightarrow{HAO} HNO_2 + 2Cyt\ c(Fe^{2+}) + 2H^+ \qquad (3.4)$$

[HAO, hydroxylamine oxidoreductase; Cyt $c(Fe^{3+})$, ferricytochrome *c*; Cyt $c(Fe^{2+})$, ferrocytochrome *c*]

Indeed, it has been reported by Arciero et al. (1991) that the enzymatic reduction rate of ferricytochrome *c*-554 with hydroxylamine is 10 times as fast as that of the ferricytochrome with “NOH.” Moreover, it has been confirmed by Andersson and Hooper (1983) that the enzymatic oxidation of hydroxylamine to nitrite proceeds in two steps as mentioned above.

In retrospect, it is recognized by the detailed analyses that when horse ferricytochrome *c* is used as the electron acceptor, nitrite of approximately one fourth as much as horse cytochrome *c* reduced is usually formed in the oxidation of hydroxylamine catalyzed by hydroxylamine oxidoreductase regardless of the amount of the ferricytochrome *c* added, if the reactions are performed in 0.1 M phosphate buffer, pH 8.0. As the rate of the reaction (3.4) is not so much slower than that of the reaction (3.3) as assumed previously, the reaction (3.4) seems to occur at a comparable rate to the reaction (3.3).

An appreciable amount of nitrite is formed, therefore, by the catalysis of hydroxylamine oxidoreductase even in the presence of less ferricytochrome *c* than its amount four times as much as hydroxylamine. However, when ferricyanide is used as the electron acceptor for the oxidoreductase, the first step of the reaction in the dehydrogenation of hydroxylamine seems to be much faster than the second step: the amount of nitrite enzymatically formed in the case where the ratio of hydroxylamine to ferricyanide is 100 μM to 100 μM is much less than that formed in the case where the ratio is 10 μM to 100 μM. Incidentally, in the enzymatic oxidation of hydroxylamine in 0.1 M glycine–NaOH buffer, pH 9.8, nitrite formed is approximately one third as much as the compound formed in the reactions performed in 0.1 M phosphate buffer, pH 8.0. Glycine may react with “NOH” to form e.g. $N_2$, and the amount of nitrite formed is diminished.

In the enzymatic oxidation of hydroxylamine catalyzed by hydroxylamine oxidoreductase, the electron acceptor for the oxidoreductase, cytochrome *c*-554, should be kept in the oxidized form as much as possible to accept electrons rapidly from hydroxylamine and “NOH.” For this purpose, sufficient air should be supplied for the bacteria to oxidize ammonia efficiently. If the air supply is not enough to oxidize hydroxylamine to nitrite, nitrous oxide ($N_2O$) occurs during the bacterial oxidation of ammonia (Poth, 1986; Anderson et al., 1993). Probably

"NOH" is changed to nitrous oxide under oxygen-deficient conditions (Hooper and Terry, 1979), though the reduction of nitrite (or nitrous acid) once fromed to the gas may also occur (Poth and Focht, 1985; Yoshida, 1988). Indeed, the bacterium has a copper protein-type nitrite reductase which catalyzes the reduction of nitrite to nitric oxide and/or nitrous oxide (Hooper, 1968; Ritchie and Nicholas, 1972, 1974; Miller and Wood, 1983; DiSpirito et al., 1985). However, as Beaumont et al. (2002) have recently found that nitrous oxide is produced even by *N. europaea* in which the DNA encoding nitrite reductase has been destroyed, the formation of nitrous oxide by the bacterium seems attributable to the decomposition of "NOH." Cantera and Stein (2007) have recently reported similar results about the formation of nitrous oxide by the nitrite reductase-deficient mutant of the bacterium.

Although *N. europaea* needs enough oxygen supply for the oxidation of hydroxylamine to nitrite under usual conditions, some of the ammonia-oxidizing bacteria appear not to need molecular oxygen for their growth under certain conditions, as will be mentioned below. As mentioned above, electrons liberated by the enzymatic "dehydrogenation" of hydroxylamine in *N. europaea* should be finally oxidized with molecular oxygen. Electrons taken out from hydroxylamine by hydroxylamine oxidoreductase are first accepted by cytochrome *c*-554 and then transferred to molecular oxygen through the electron transfer system composed of cytochromes:

$$NH_2OH \rightarrow \underset{\text{reductase}}{NH_2OH \text{ oxido-}} \rightarrow \text{Cyt } c\text{-554} \rightarrow \text{Cyt } c\text{-552} \rightarrow \underset{\substack{(=\text{Cyt } c\\ \text{oxidase})}}{\text{Cyt } aa_3} \rightarrow O_2$$

As hydroxylamine oxidoreductase has been described above, some properties of other components involved in the above electron transfer system will be described below.

### (b) Cytochrome *c*-554

Cytochrome *c*-554 was first isolated by Yamanaka and Shinra (1974). The cytochrome has four heme C molecules in the molecule with molecular mass of 25 kDa (Andersson et al., 1986). Its isoelectric point is at pH 10 and functions as the electron acceptor for hydroxylamine oxidoreductase (Yamanaka and Shinra, 1974). The four heme C molecules of cytochrome *c*-554 stack in parallel with each other, and the 5th and 6th axial ligands to each of three of the four heme C molecules are histidine residues, while the remaining one heme molecule is ligated with one histidine residue at the 5th ligating site, and its 6th site is open (Iverson et al., 2001). Thus, the cytochrome is autoxidizable, but strangely enough, its reduction rate with hydroxylamine catalyzed by hydroxylamine oxidoreductase hardly varies with the absence or presence of molecular oxygen (Yamanaka and Shinra, 1974). Recently, it has been found that cytochrome *c*-554 has NO reductase activity (Upadhyay et al., 2006). The amino acid sequence of cytochrome *c*-554 has been deduced from DNA sequence which encodes the cytochrome (Bergmann et al., 1994).

### (c) Cytochrome *c*-552

Cytochrome *c*-552 is a monoheme cytochrome *c* with molecular mass of 9.3 kDa. The structure of the cytochrome has been supposed to be similar to that of *Pseudomonas aeruginosa* cytochrome *c*-551 on the basis its reactivity with several oxidoreductase (Yamanaka and Shinra, 1974). Indeed, its amino acid sequence is very similar to that of *P. aeruginosa* cytochrome *c*-551 (Fujiwara et al., 1995; Timkovich et al., 1998). Although cytochrome *c*-552 does not act as the electron acceptor for hydroxylamine oxidoreductase, it is reduced with hydroxylamine by the catalysis of the enzyme in the presence of a catalytic amount of cytochrome *c*-554. Furthermore, ferrocytochrome *c*-552 is oxidized with molecular oxygen by the catalysis of cytochrome *c* oxidase (cytochrome $aa_3$) as described below (Yamazaki et al., 1985).

Cytochrome *c*-552 belongs to cytochrome $c_6$ or $c_8$ group. As the cytochromes of this group react rapidly with *Pseudomonas aeruginosa* nitrite reductase but does not react with cow cytochrome *c* oxidase (Yamanaka, 1992), their structures have been supposed to be similar to each other. Thus, their amino acid sequences resemble each other though the sequence of *Chlorobium limicola* f. *thiosulfatophilum* cytochrome *c*-555 shows smaller similarity to other cytochromes $c_6$ or $c_8$ than the similarities that other cytochromes $c_6$ or $c_8$ show each other (Table 3.2).

### (d) Cytochrome *c* Oxidase

Cytochrome *c* oxidase of *N. europaea* was called cytochrome $a_1$, as it shows the α peak at 595 nm (Erickson et al., 1972). However, the electrophoretically homogeneous preparation of the oxidase has two heme A molecules and two copper atoms (CuA and CuB) in the molecule, and one of the two heme A molecules reacts with

**Table 3.2.** Comparison in amino acid sequence of *N. europaea* cytochrome *c*-552 with several cytochromes $c_6$ or $c_8$

| Cytochrome *c* | α peak (nm) | Number of identical residues | | | | |
|---|---|---|---|---|---|---|
| | | (1) | (2) | (3) | (4) | (5) |
| (1) *N. europaea* | 552 | 81 | | | | |
| (2) *P. aeruginosa* | 551 | 45 | 82 | | | |
| (3) *P. mendocina* | 551 | 48 | 49 | 82 | | |
| (4) *P. stutzeri* | 552 | 46 | 55 | 63 | 82 | |
| (5) *C. limicola* f. *thiosulfatophilum* | 555 | 19 | 18 | 18 | 18 | 86 |

(1) Fujiwara et al. (1995), Timkovich et al. (1998); (2) Ambler (1963); (3), (4) Ambler and Wynn (1973); (5) Van Beeumen and Ambler (1973)

carbon monoxide. Namely, the oxidase shows many properties of cytochrome $aa_3$ (Yamazaki et al., 1985; DiSpirito et al., 1986). Therefore, the oxidase should be called cytochrome $aa_3$ but not cytochrome $a_1$.

The researchers of cytochrome tend to call the cytochrome showing the α peak at the wavelengths shorter than 600 nm cytochrome $a_1$, though Keikin and Hartree (1939) defined cytochrome $a_1$ as the cytochrome having the α peak around 590 nm (shorter than 595 nm!). Now, the difference between cytochrome $a_1$ and cytochrome $aa_3$ should be decided according to the molecular structures and enzymatic activity in addition to the position of the α peak. On the basis of this criterion, probably all previous cytochromes $a_1$ like *N. europaea* and *Acidithiobacillus ferrooxidans* (see below) cytochromes $a_1$ belong to cytochrome $aa_3$. Only two cytochromes $a_1$ are known at present: cytochrome $a_1c_1$ of *Nitrobacter winogradskyi* (see below) and cytochrome $a_1a$ of *Sulfolobus acidocaldarius*. These cytochromes do not catalyze the reduction of molecular oxygen unlike the previous description by Morton (1958). Cytochrome $a_1c_1$ is a nitrite oxidoreductase (Tanaka et al., 1983) and cytochrome $a_1a$ is a component of quinol dehydrogenase (Lübben et al., 1992).

Although the activity of purified *N. europaea* cytochrome *c* oxidase to catalyze the oxidation of *N. europaea* ferrocytochrome *c*-552 is fairly low in vitro, it is much stimulated by addition of poly-L-lysine. For example, one molecule of the oxidase catalyzes the oxidation of nine molecules of ferrocytochrome *c*-552 per second in 40 mM phosphate buffer, pH 6.0, while the activity is raised up to the oxidation of 64 molecules of the ferrocytochrome *c* per second on addition of 5 μM poly-L-lysine (Yamazaki et al., 1988). Therefore, poly-L-lysine or similar compounds may occur in the plasma membrane of the bacterium, or some similar microenvironments caused by the presence of poly-L-lysine may occur in vivo.

The affinity to carbon monoxide (CO) of the oxidase is very low as compared to other cytochromes *c* oxidase; though the oxidase binds completely with carbon monoxide in 100% carbon monoxide atmosphere, its CO-complex dissociates easily to the oxidase and carbon monoxide in air unlike other cytochromes *c* oxidase (Erickson et al., 1972; Yamazaki et al., 1985) except for *Starkeya novella* cytochrome *c* oxidase monomer, which does not bind with carbon monoxide bubbled in air (Shoji et al., 1992).

*Nitrosomas europaea* cytochrome *c* oxidase purified from the bacterial cells cultured in a copper-deficient medium has 1 copper atom per 2 molecules of heme A. As the copper-deficient oxidase does not show $g = 2.0$ signal in EPR spectrum, it seems not to have CuA. However, the copper-deficient oxidase catalyze the oxidation of horse ferrocytochrome *c* at the same rate as the two- (or three)-copper containing oxidase does, though its activity to catalyze the oxidation of *N. europaea* ferrocytochrome *c*-552 is one third of that of the two- (or three)-copper oxidase (Numata et al., 1989). These results suggest that cytochrome *c* oxidase needs not necessarily CuA to catalyze the oxidation of ferrocytochrome *c*. Thus, from *Rhodobacter sphaeroides*, cytochrome $cbb_3$ has been obtained, which does not have CuA and still shows cytochrome *c* oxidase activity (Garcia-Horsman et al., 1994).

### *3.1.3 Electron Transfer Pathway Coupled to the Oxidation of Ammonia*

As mentioned above, ammonia is oxidized to nitrous acid via hydroxylamine in *N. europaea*; first ammonia is oxidized to hydroxylamine by the catalysis of ammonia monooxygenase, and hydroxylamine formed is oxidized to nitrous acid by the catalysis of hydroxylamine oxidoreductase. Molecular oxygen is not necessary to the reaction itself of $NH_2OH \rightarrow HNO_2$ (Yamanaka and Sakano, 1980) but it is required for the consumption of electrons liberated from the reaction, $NH_2OH + H_2O \rightarrow HNO_2 + 4H^+ + 4e$. Electrons thus liberated are transferred first to cytochrome *c*-554, then to cytochrome *c*-552, and finally oxidized with molecular oxygen by the catalysis of cytochrome *c* oxidase. Based on the results described above, the electron transfer pathway in the oxidation of ammonia to nitrite or nitrous acid by *N. europaea* will be presented as shown in Fig. 3.3.

Cytochrome *b*-560 has been obtained from the bacterium (Fukumori et al., 1988a), and the occurrence of two membrane-bound *b*-type cytochromes has been observed (Miller and Wood, 1983). As no cytochrome is involved in the system of the oxidation of ammonia and hydroxylamine, the *b*-type cytochrome(s) will function in the reduction system of $NAD(P)^+$ in which ubiquinone-8 seems to be involved (Hooper et al., 1972). Although an electron transfer pathway in which ubiquinone-8 mediates electrons between cytochrome *c*-554 and cytochrome *c*-552

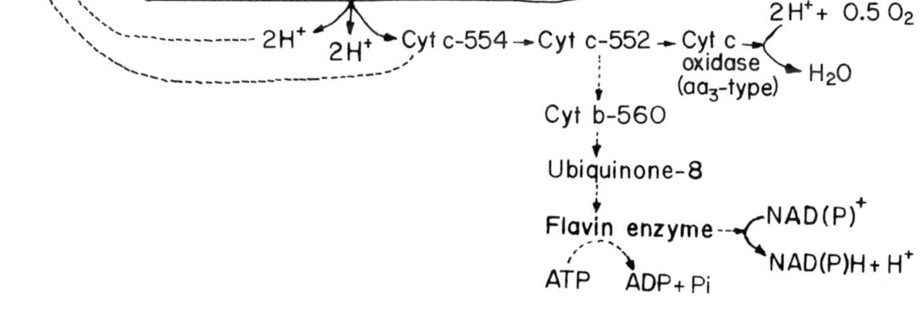


**Fig. 3.3.** A scheme presenting the oxidation mechanism of ammonia to nitrous acid by *Nitrosomonas europaea* (prepared mainly on the basis of Lees, 1952; Aleem, 1966; Yamanaka and Shinra, 1974; Yamanaka and Sakano, 1980; Suzuki and Kwok, 1981; Andersson and Hooper, 1983; Yamazaki et al., 1985; Igarashi et al., 1997). *Dashes with arrows*, unverified; *Cyt*, cytochrome; *Pi*, phosphate

was proposed (Wood, 1986), no experimental evidence has been uncovered regarding the system. Moreover, it has been reported by McTavish et al. (1995) that the electrons are not transferred directly from cytochrome *c*-554 to ubiquinone-8.

It seems very difficult to reconstitute the hydroxylamine oxidase system by mixing purified components, because the optimal pHs of hydroxylamine oxidoreductase and cytochrome *c* oxidase are 9.6 and 5.6, respectively. However, as the optimal pH of the oxidoreductase is lowered to around 7 in the presence of ubiquinone-8 and oleic acid, the reconstitution of the system seems possible (Numata, 1989). When hydroxylamine and the oxidoreductase are added to the mixture of cytochromes *c*-554 and *c*-552 supplemented with ubiquinone-8 and oleic acid, cytochrome *c*-552 is reduced at pH 7.0, and ferrocytochrome *c*-552 thus formed is oxidized on addition of cytochrome *c* oxidase. Although these reactions are spectrally observed, the formation of nitrite is not observed, unfortunately, in the above system.

The ammonia-oxidizing bacteria biosynthesizes the cellular materials from carbon dioxide. For this purpose, they need NAD(P)H. Electrons to reduce $NAD(P)^+$ seem to come from ferrocytochrome *c*-552 by the supply of energy, because Aleem (1966) reported that he had demonstrated that $NAD(P)^+$ was anaerobically reduced with horse ferrocytochrome *c* on addition of ATP using the cell-free extracts of *N. europaea*, though the enzymatic system participating in the reduction of $NAD(P)^+$ has not been known. However, every attempt by the author and his colleagues to reproduce his results has been unsuccessful to date.

### *3.1.4 Dehalogenation of Chloroethylenes by Bacteria*

Tetrachloroethylene and trichloroethylene are used for the cleaning of metals finely purified and for dry cleaning of clothes as the cleaning solvents. As they are very poisonous we have to take care not to pollute the environment with these compounds. When we accidentally litter the environment with these compounds, the cleaning methods which remove them effectively using bacteria may be helpful.

*Nitrosomonas europaea* dehalogenates trichloroethylene. Ammonia monooxygenase of the bacterium is responsible for the dehalogenation. However, as the enzyme is destroyed by hydrogen chloride resulting from the dehalogenation, the dehalogenation by the bacterium does not continue for very long (Hyman et al., 1995). The dehalogenation activity of the bacterium is recovered if trichloroethylene is removed from the bacterium when the bacterial activity has come down. The bacterium does not attack tetrachloroethylene (Arciero et al., 1989).

Some bacteria have been found that dehalogenate tetrachloroethylene (Maymo-Gatell et al, 1997; Holliger et al., 1999). For example, *Dehalococcoides ethenogenes* anaerobically dehalogenates tetrachloroethylene with hydrogen gas in a stepwise manner; tetrachloroethylene [perchloroethylene (PCE)] is changed suc-

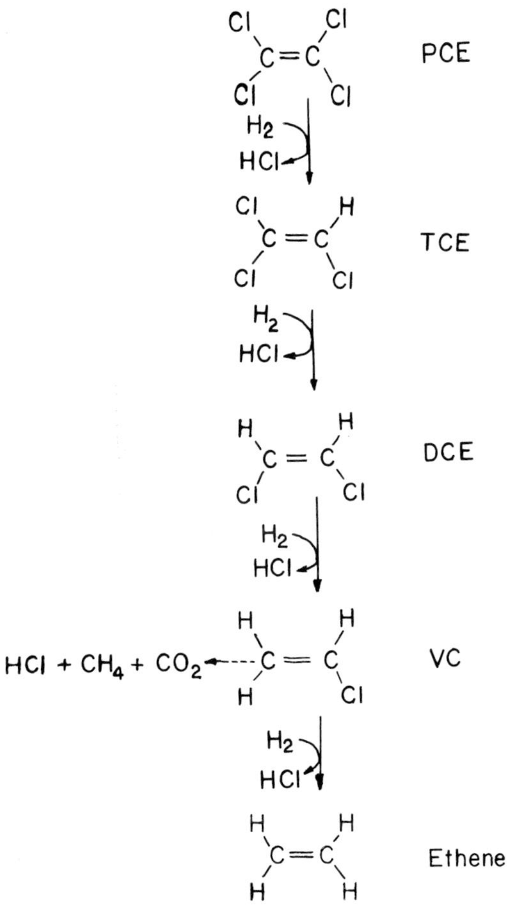


**Fig. 3.4.** The reaction steps of bacterial dehalogenation (prepared on the basis of Maymo-Gatell et al., 1979; Uchiyama, 1999). *PCE*, perchloroethylene; *TCE*, trichloroethylene; *DCE*, dichloroethylene; *VC*, vinyl chloride

cessively to trichloroethylene (TCE), dichloroethylene (DCE), vinyl chloride (VC), and ethene (ethylene). Among the processes, the final step is fairly slow. This step is a bottleneck in the dehalogenation of tetrachloroethylene by the bacterium (Fig. 3.4).

The dehalogenating bacteria anaerobically oxidize hydrogen gas with tetrachloroethylene to produce ATP; the dehalogenation by the bacteria is a respiration, namely, dehalogenation respiration. It may sound marvelous that the dehalogenation is a respiration in which the bacteria oxidize hydrogen with tetrachloroethylene and its derivatives. However, the enzymatic reactions which seem to participate in respiratory processes have been found in the bacteria; tetrachloroethylene is reduced with hydrogen by the catalysis of tetrachloroethylene reductive dehalogenase and hydrogenase catalyzes the reduction of menaquinone with hydrogen gas (Holliger et al., 1999). In any case, it seems possible for us to clean the soil polluted by tetrachloroethylene or trichloroethylene using the bacteria. This is known as bioremediation to cleanse the pollution caused by harmful substances using microorganisms.

### 3.1.5 Various Growth Features of Ammonia-Oxidizing Bacteria

Although the ammonia-oxidizing bacteria were previously thought to grow under strict lithoautotrophic conditions and strict aerobic conditions, the growth of some bacteria of *Nitrosomonas* genus has been now known to be stimulated by organic compounds and to occur anaerobically. The growth of *Nitrosomonas europaea* is stimulated by addition to culture medium of the compounds in citrate cycle (Kamiyoshi and Tabata, 1991). On the contrary, it is also reported that a considerable part of ammonia in the medium for the bacterium is not oxidized to nitrite but is changed to nitric oxide and nitrous oxide in the presence of pyruvate, yeast extracts, or bactopeptone (Stüven et al., 1992). The bacterium forms nitrous oxide under oxygen-deficient conditions, as already mentioned (Poth, 1986) and the formation of the compound is stimulated by the presence of pyruvate. Moreover, *Nitrosomonas* sp. has been found which forms nitrogen gas under anaerobic conditions (Poth, 1986). The ammonia-oxidizing bacteria are roughly classified into three groups with regard to their utilization of organic compounds; the growth of the first group of bacteria is stimulated by organic compounds, that of the second group is insensitive to the compounds, and that of the third group is inhibited by the compounds. However, no direct relation is observed between the stimulation by the organic compounds of nitrite formation and the increase in assimilation of the compounds (Krummel and Harms, 1982).

*Nitrosomas europaea* does not grow on hydroxylamine, but does grows on the compound in the presence of ammonium ion (molar growth yield for hydroxylamine, $Y_{NH2OH} = 4.74$) (de Bruijin et al., 1995). The reason is not known why ammonium ion is necessary for the bacterium to grow on hydroxylamine. Furthermore, the bacterium anaerobically oxidizes hydroxylamine with nitrite to form nitrous oxide.

*Nitrosomonas eutropha* reduces nitrite or nitrogen dioxide with ammonia or hydroxylamine. This reaction is thought to proceed through nitric oxide as the intermediate metabolite and to supply the energy for the bacterium to survive under anaerobic conditions. However, the bacterium does not grow by the reaction (Jetten et al., 1999). It anaerobically oxidizes ammonia to nitrite and nitric oxide in the presence of nitrogen dioxide, and simultaneously reduces 40%–60% of nitrite formed to nitrogen gas. The bacterium grows on the above reactions when carbon dioxide is utilizable (Schmidt and Bock, 1997). The oxidation of ammonia to nitrite mentioned above proceeds through the use of hydroxylamine as intermediate metabolite. The mechanism of this anaerobic oxidation of ammonia is interesting. Moreover, it seems mysterious that the bacterium grows in the presence of higher concentrations of nitrogen dioxide, because the compound is very toxic for organisms. The anaerobic oxidation of ammonia to nitrite with nitrogen dioxide or dinitrogen tetraoxide ($N_2O_4$) through hydroxylamine as the intermediate metabolite is also found with the cell-free extracts of the bacterium (Schmidt and Bock, 1998). The aerobic oxidation of ammonia by *N. eutropha* is found to be stimulated in the presence of nitric oxide (Zart et al., 2000).

*Nitrosomonas europaea* anaerobically fixes carbon dioxide consuming nitrite in the presence of ammonia and pyruvate (Abeliovich and Vonshak, 1992). Furthermore, *N. eutropha* anaerobically grows by the oxidation of hydrogen gas with nitrite at the redox potential of –0.25 to –0.3.0 V (Bock et al., 1995). It is found by the studies on the biosynthesis of ammonia monooxygenase that *N. europaea* can acquire energy for the growth by the oxidation of pyruvate and hydrazine with nitrite under the anaerobic conditions (Hyman and Arp, 1995). In this case, most of the nitrite is reduced to nitrogen gas, while a small amount of nitrous oxide is formed. The bacterium also anaerobically catalyzes the reaction,

$$NH_4^+ + NO_2^- \rightarrow N_2 + 2H_2O.$$

However, the bacterium does not grow on this reaction (Bock et al., 1995).

## *3.1.6 Bacterial Oxidation of Nitrite*

As mentioned above, ammonia is oxidized to nitrite or nitrous acid by the ammonia-oxidizing bacteria. As nitrite is poisonous for most organisms, it is utilized neither by animals nor plants, though some bacteria utilize it. In Nature, the nitrite-oxidizing bacteria are present together with the ammonia-oxidizing bacteria, and nitrite formed by the latter bacteria is quickly oxidized to nitrate by the former. When nitric acid is formed, it usually reacts quickly with calcium carbonate to form calcium nitrate in the soil. Nitrates are a good nitrogen source for all plants.

Many nitrite-oxidizing bacteria have been discovered, for example *Nitrobacter winogradskyi*, *Nitrobacter hamburgensis*, *Nitrospina gracilis*, *Nitrococcus mobilis*, and *Nitrospira marina*, (Watson et al., 1989). Among them, *N. winogradskyi* is studied most extensively.

When the nitrite-oxidizing bacteria oxidize nitrite to nitrate, oxygen from water but not from molecular oxygen is incorporated (Aleem et al., 1965). Therefore, two of three oxygen atoms in nitrate formed from ammonia by the actions of the ammonia-oxidizing and nitrite-oxidizing bacteria come from water. However, as mentioned below, in the case of nitrate formed by a certain heterotrophic nitrifier, two of the three oxygen atoms of nitrate come from molecular oxygen.

### (a) Nitrite Oxidoreductase

Nitrite oxidoreductase which participates in the first step of the oxidation of nitrite in *Nitrobacter winogradskyi* (formerly, *N. agilis*) was found to show an absorption peak at 587 nm (Lees and Simpson, 1957). Therefore, the enzyme was thought to be cytochrome $a_1$ (Aleem, 1970, 1977). When the enzyme is purified to an electrophoretically homogeneous state, it is found to have heme C besides heme A. So the enzyme is called also cytochrome $a_1c_1$ (Tanaka et al., 1983).

The enzyme contains two heme A molecules, two heme C molecules, one molybdenum atom, and five $[Fe_4S_4]$ clusters in the molecule with molecular mass of 250 kDa. Molybdenum occurs as a complex with molybdopterin guanine dinucleotide (MGD) (Fukuoka et al., 1987; Yoshino, 1994; Suzuki et al., 1997). The enzyme catalyzes the reduction of *N. winogradskyi* ferricytochrome *c*-550 and horse ferricytochrome *c* with nitrite, i.e., it catalyzes the oxidation of nitrite with ferricytochromes *c* as the electron acceptor. The reaction is stimulated by $Mn^{2+}$ and $Ca^{2+}$. Although the enzyme catalyzes actively the oxidation of nitrite around pH 8, it shows also capability to catalyze the reduction of nitrate with ferrocytochrome *c* at pHs less than 6; the bacterium changes itself from a nitrifier to a "denitrifier" at pHs less than 6.

Nitrite oxidoreductase has been purified also from *Nitrobacter hamburgensis* (Sundermeyer-Klinger et al., 1984) and *Nitrospira moscoviensis* (Spieck et al., 1998). The *N. hamburgensis* enzyme has been reported to have heme C but not to have heme A. It does not catalyze the reduction of horse ferricytochrome *c* with nitrite although it catalyzes the reduction of ferricyanide. The *N. moscoviensis* enzyme has been reported to have heme B but not to have either heme A or heme C. Although the *N. moscoviensis* enzyme catalyzes the oxidation of nitrite with chlorate as the electron acceptor, it has not been reported whether cytochrome *c* or ferricyanide acts as the electron acceptor for the enzyme. The enzyme of this bacterium is known to be located in the periplasmic space. Table 3.3 lists some properties of nitrite oxidoreductase purified from three species of the nitrite-oxidizing bacteria.

The *N. winogradskyi* enzyme removes electrons from nitrite plus water, and gives the electrons to cytochrome *c*-550. The electrons accepted by the cytochrome are finally given to molecular oxygen by the catalysis of cytochrome *c* oxidase.

$$NO_2^- + H_2O + 2\text{Cyt } c\text{-550}(Fe^{3+}) \xrightarrow{\text{Nitrite oxidoreductase}} NO_3^- + 2\text{Cyt } c\text{-550}(Fe^{2+}) + 2H^+ \quad (3.5)$$

$$2\text{Cyt } c\text{-550}(Fe^{2+}) + 2H^+ + 0.5O_2 \xrightarrow{\text{Cytochrome } c \text{ oxidase}} 2\text{Cyt } c\text{-550}(Fe^{3+}) + H_2O \quad (3.6)$$

[Cyt *c*-550($Fe^{3+}$) and Cyt *c*-550($Fe^{2+}$) are ferricytochrome *c*-550 and ferrocytochrome *c*-550, respectively]

In the processes of the enzymatic oxidation of nitrite, the reduction by nitrite of cytochrome *c*-550 hardly occurs because $E_{m,7.0}$ (midpoint redox potential) of nitrite/nitrate system (+0.40 V) is higher than that of cytochrome *c*-550 (+0.28 V). The reduction step of molecular oxygen catalyzed by cytochrome *c* oxidase proceeds rapidly, because $E_{m,7.0}$ of $H_2O/O_2$ system is +0.82 V and is very much higher than that of cytochrome *c*-550. Therefore, in the bacterium, the electron flow from nitrite

**Table 3.3.** Comparison of nitrite oxidoreductases purified from three species of the nitrite-oxidizing bacteria

| Properties | Bacteria | | |
|---|---|---|---|
| | *N. winogradskyi* (1) | *N. hamburgensis* (2) | *N. moscoviensis* (3) |
| Molecular mass (kDa) | | | |
| Active enzyme | 254 | 394 | 267 |
| Subunit | 145 | 116 × 2 | 130 |
| | 59 | 65 × 2 | 62 |
| | 30 | 32 | 46 |
| | 20 | | 29 |
| Heme (molecule/molecule) | 2A, 2C | C | B |
| Metals (atom/molecule) | | | |
| Mo (molybdenum factor) | 1 | 0.7 | + |
| Fe/S cluster | $5[Fe_4S_4]$ | 23.0 (total Fe)[a] | ND[b] |
| Mn | 0.18 | ND | ND |
| Cu | 0.13 | 0.89 | ND |
| Zn | ND | 1.76 | ND |
| Oxidation of nitrite (molecular activity/min) | 9.12 | ND | ND |
| Optimal pH | ~8 | 8.0 | ND |
| $K_m$ (mM) | | | |
| for nitrite | 0.63 | 3.6 | ND |
| for nitrate | ND | 0.9 | ND |
| for horse cytochrome *c* | 0.0061 | No reaction | ND |
| Inhibitor and its concentration for 50% inhibition ($\mu$M) | | | |
| Cyanide | ~8 | ND | ND |
| Azide | ~8 | ND | ND |
| Nitrate | 440 | ND | ND |
| Location in the cell | ND | ND | Periplasm |

(1) Tanaka et al. (1983), Fukuoka et al. (1987), Yoshino (1994), Suzuki et al. (1997); (2) Sundermeyer-Klinger et al. (1984), Meincke et al. (1992); (3) Spieck et al. (1998)

[a] S content is not reported

[b] Not determined

is propelled by the reduction of molecular oxygen; the whole electron transfer from nitrite to oxygen, in vivo, proceeds easily. Previously, the potential of the cytoplasm membrane was thought to be necessary for the reduction step of cytochrome *c*-550 with nitrite (Cobley, 1976), because the ionophore diminished the nitrite oxidation rate by *Nitrobacter winogradskyi* (Cobley, 1984). However, even if the concentration of the ionophore was increased, the inhibition degree of the oxidation rate of nitrite by the bacterium was at most 50%. So it was doubtful whether or not the membrane potential is inevitable for the reduction of the cytochrome with nitrite. Nomoto et al. (1993) have demonstrated with the reconstituted nitrite oxidase system that the membrane potential is unnecessary for the oxidation of nitrite in the bacterium, as will be mentioned below. The author and his coworkers purified from the bacterium all the components that seemed to participate in nitrite oxidation; cytochrome $a_1c_1$ (nitrite oxidoreductase) (Tanaka et al., 1983), soluble cytochrome *c*-550 [cytochrome *c*550(s)] (Ketchum et al., 1969; Yamanaka et al., 1982), membrane bound cytochrome *c*-550 [cytochrome *c*-550(m)] (Nomoto et al., 1993), and cytochrome $aa_3$ (cytochrome *c* oxidase) (Yamanaka et al., 1979a, 1981a). The properties of the components will be briefly described below except for nitrite oxidoreductase (cytochrome $a_1c_1$) which has been already mentioned in this section.

### (b) Cytochromes *c*-550(s) and *c*-550(m)

Cytochrome *c*-550(s) was partially purified by Ketchum et al. (1969). Afterward it was purified to an electrophoretically homogeneous state (Yamanaka et al., 1982), and its complete amino acid sequence was determined (Tanaka et al., 1982). Its molecular mass is 12.4 kDa. The cytochrome is very similar to mitochondrial cytochrome *c* (similarity, 40%–49%) on the basis of the sequence, and reacts with yeast cytochrome *c* peroxidase at the rate of 79% as fast as mitochondrial cytochrome *c*. Ferrocytochrome *c*-550(s) is oxidized very fast with molecular oxygen by the catalysis of *N. winogradskyi* cytochrome *c* oxidase; turnover number is $117\ s^{-1}$ (Yamanaka et al., 1982; Nomoto et al., 1993).

Another cytochrome *c*, cytochrome *c*-550(m), was purified from the bacterium with the aid of the detergent, Triton X-100 (Nomoto et al., 1993). This cytochrome spectrally resembles cytochrome *c*-550(s), while the N-terminal amino acid sequence differs between them. Its molecular mass is 13.6 kDa.

### (c) Cytochrome *c* Oxidase

The oxidase is very similar to mitochondrial cytochrome *c* oxidase in spectral properties. It has two heme A molecules and two copper atoms in the molecule, and one of the two heme A molecules reacts with carbon monoxide. Thus, it is cytochrome $aa_3$. However, its molecule is composed of only two subunits (40 kDa and 27 kDa) (Yamanaka et al., 1979a, 1981a) unlike the mitochondrial oxidase comprising 13 subunits (Kadenbach, 1983; Tsukihara et al., 1995). Although the presence of a third subunit is indicated (Chaudhry et al., 1980; Berben, 1996), it is true that the oxidase composed of the two subunits mentioned above shows cytochrome *c* oxidase activity. The oxidase catalyzes the fast oxidation of both ferrocytochrome *c*-550(s) and ferrocytochrome *c*-550(m), though the enzymatic oxidation of the former cytochrome is much faster than the latter (Nomoto et al., 1993). However, the oxidase does not show any proton pumping activity (Sone et al., 1983) as will be described below. This cannot be attributable to the absence of the third subunit, because the proteoliposomes in which the membrane fractions of *N. winogradskyi* are reconstituted do not show the activity either (Sone, 1986).

### (d) Reconstitution of Nitrite Oxidation System

The author and his colleagues succeeded in the reconstitution of a nitrite oxidase system using the purified components as mentioned above [cytochrome *c*-550(m), cytochrome $a_1c_1$ and cytochrome $aa_3$]. When cytochrome *c*-550(m) was added to the proteoliposomes containing cytochrome $a_1c_1$, cytochrome $aa_3$, and nitrite, an instant oxygen consumption was observed (Nomoto et al., 1993). The oxygen consumption rate per cytochrome $a_1c_1$ was about 16% of that of the membrane fractions prepared from *N. winogradskyi*. When cytochrome *c*-550(s) was used in place of cytochrome *c*-550(m), an appreciable oxygen consumption was not observed. This may be attributable to the low affinity of cytochrome *c*-550(s) to the proteoliposomes. Therefore, the results obtained do not mean that cytochrome *c*-550(s) does not function in the bacterial cells in the oxidation of nitrite. Rather, cytochrome *c*-550(s) seems to function in the nitrite oxidase system in vivo as well as cytochrome *c*-550(m), because the amounts in the cells of both the cytochromes are comparable. Furthermore, it is expected that cytochrome *c*-550(s) is a better electron mediator in the nitrite oxidase system in vivo than cytochrome *c*-550(m), because the reactivity of the former cytochrome with cytochrome *c* oxidase is much higher than that of the latter (Nomoto et al., 1993). In any case, the nitrite-oxidizing system of the bacterium is composed of the three or four components mentioned above. In particular, cytochrome *b* is unnecessary for the system. Furthermore, the results obtained above mean that the potential of the cytoplasm membrane is not necessary for the bacterial oxidation of nitrite. Besides the enzymes and cytochromes mentioned above, cytochrome *b*-559 (Kurokawa et al., 1989) and a NAD(P)H-cytochrome *c* oxidoreductase (Kurokawa et al., 1987) are purified from the bacterium. As the oxidoreductase catalyzes the reduction of $NAD(P)^+$ with benzylviologen radical, it seems probable that the enzyme functions as a NAD(P)-reductase, and cytochrome *b*-559 may donate electrons to the enzyme. As $E_{m,7.0}$ of $NAD(P)H/NAD(P)^+$ system is –0.32 V, while that of the $NO_2^-/NO_3^-$ system is about +0.4 V, the reduction of $NAD(P)^+$ by nitrite requires energy. Indeed, Aleem et al. (1963) showed with the cell-free extracts of *Nitrobacter agilis* (now *N. winogradskyi*) that anaerobic reduction of $NAD(P)^+$ with ferrocytochrome *c* occurred on addition of ATP. On the basis of the results mentioned herein, the electron transfer pathways coupled to the oxidation of nitrite by *N. winogradskyi* are illustrated in Fig. 3.5.

However, no one has succeeded in the reconfirmation of the results obtained by Aleem et al. (1963) on the reduction of $NAD(P)^+$ with ferrocytochrome *c*. Freitag and Bock (1990) have reported that as the reduction of $NAD^+$ with nitrite in *Nitrobacter* is not inhibited by DCCD which inhibits $F_0F_1$-ATPase, i.e., inhibits energy transfer through cytoplasm membrane (in the case of bacteria), the reduction of $NAD^+$ is not performed by using the energy of ATP. They claim that nitric oxide participates in the formation of NADH, and the ratio of nitric oxide consumed/

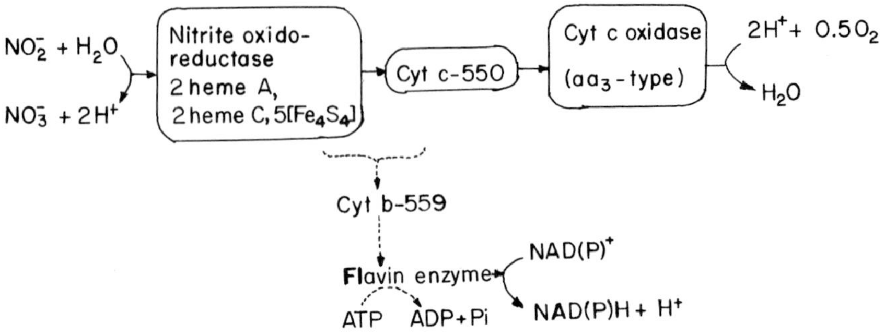


**Fig. 3.5.** The electron transfer pathways proposed for the oxidation of nitrite by *Nitrobacter winogradskyi* (prepared mainly on the basis of Aleem et al., 1963, 1965; Yamanaka et al., 1981a; Tanaka et al., 1983; Nomoto et al., 1993). *Dashes with arrows*, unverified; *Cyt*, cytochrome; *Pi*, phosphate

NADH formed is 2.3. However, the mechanism by which $NAD^+$ is reduced by nitric oxide is not as yet known.

Although it is doubtless that the nitrite-oxidizing bacteria biosynthesize ATP coupled to the oxidation of nitrite as the bacteria grow on nitrite, there is a mystery about the mechanisms by which ATP is biosynthesized in the bacteria. The proton pumping activity of *N. winogradskyi* cytochrome *c* oxidase is not observed as mentioned above (Sone et al., 1983). Furthermore, the bacterial spheroplasts do not show either discharge or intake of protons when they oxidize nitrite (Hollocher et al., 1982). Although it has been claimed that the proton translocation activity is observed with the spheroplasts of *N. winogradskyi* using isoascorbate plus *N,N,N′,N′*-tetramethyl-*p*-phenylenediamine as the electron donor (Wetzstein and Ferguson, 1985), it may be caused by the dissociation of isoascorbate; protons left behind when electrons are donated from isoascorbate to the oxidase will come out from the spheroplasts. Indeed, the proteoliposomes in which the membrane fractions of *N. winogradskyi* are incorporated do not show proton pumping activity on ferrocytochrome *c* pulse (Sone, 1986). As $F_1$-type ATPase is obtained from the bacterium (Hara et al., 1991), the ATP biosynthesis in the bacterium is thought also to be dependent on the electrochemical potential of the proton. Freigtag and Bock (1990) have reported that ATP is biosynthesized in *Nitrobacter* by the oxidation of NADH with molecular oxygen or nitrate. However, the mechanisms have not been clarified which relate the oxidation of NADH and the biosynthesis of ATP. As nitrite oxidoreductase is found to locate in the periplasm (Spieck et al., 1998), it might be possible that protons liberated in the oxidation of nitrite by the catalysis of the oxidoreductase at the periplasm of the bacterium go to the cytoplasm through $F_0F_1$-ATPase, and are instantly oxidized with oxygen by the catalysis of cytochrome *c* oxidase at the inside of the cytoplasm membrane. Thus, although protons are formed in the enzymatic oxidation of nitrite, the pH of the culture medium is not lowered or is rather raised a little when the bacterium is cultured.

### *3.1.7 Nitrification by Heterotrophic Bacteria*

The nitrifying bacteria mentioned above are chemolithoautotrophs. The ammonia-oxidizing and nitrite-oxidizing bacteria acquire the energy necessary for the life processes, i.e., ATP, by oxidizing ammonia and nitrite, respectively. In Nature, besides these, heterotrophic nitrifying bacteria also occur which simultaneously oxidize organic compounds and ammonia (Verstraete and Alexander, 1973; Castignetti and Hollocher, 1982). The heterotrophic nitrifying bacteria can oxidize ammonia to nitrate (Rapen et al., 1989; Nishio et al., 1994). On this point, the bacteria differ from the chemolithoautotrophic nitrifiers. As the heterotrophic nitrifiers can biosynthesize ATP by the oxidation of organic compounds, it seems unnecessary for the bacteria to acquire additional ATP by the oxidation of ammonia. However, it is known that the cellular weight increases by oxidizing hydroxylamine (Jetten et al., 1997) in *Pseudomonas* PB16, and ATP is biosynthesized coupled to the oxidation of hydroxylamine in *Methylococcus thermophilus* (Malashenko et al., 1979). On the contrary, some results have been obtained which support that the nitrification in the heterotrophic nitrifiers occurs for detoxification (Ono et al., 1996). It should be clarified exactly by future studies why the heterotrophic nitrifiers perform the nitrification (Castignetti et al., 1983, 1990). In any case, the heterotrophic nitrifiers form nitrite and nitrate from ammonia. The nitrification activity per cell of the heterotrophic nitrifiers is said to be about one thousandth of that of the chemolithoautotrophic nitrifiers. However, as the global mass of the heterotrophic nitrifiers seems much larger than that of the chemolithoautotrophic nitrifiers, the total activity on Earth of the nitrification by the heterotrophic nitrifiers are said to be comparable to that by the chemolithoautotrophic nitrifiers (Catignetti and Hollocher, 1982; Hall, 1986). Therefore, the nitrification by the heterotrophic nitrifiers cannot be ignored from the viewpoint of the nitrogen circulation on Earth.

Although the oxidation mechanism of nitrite to nitrate in the heterotrophic nitrifiers has not been known at all on the enzyme level, the oxidation mechanism of ammonia to nitrite has been partially clarified. Ammonia is oxidized to nitrite through hydroxylamine also in the heterotrophic bacteria. The oxidation of ammonia to hydroxylamine is catalyzed by ammonia monooxygenase as in the enzyme of *Nitrosomonas europaea*. The enzyme purified from *Paracoccus pantotropha* GB17 (formerly *Thiosphaera pantotropha* GB17 or *Paracoccus denitrificans* GB17) catalyzes the oxidation of ammonia to hydroxylamine and contains copper, but its activity is not inhibited by acetylene (Moir et al., 1996), unlike the enzyme of *Nitrosomonas europaea*.

From several heterotrophic nitrifiers, enzymes are obtained which catalyze the reduction of ferricytochrome *c* in the presence of hydroxylamine like hydroxylamine oxidoreductase of *N. europaea* does, but they do not have heme C, unlike the *N. europaea* enzyme (Kurokawa et al., 1985; Wehrfritz et al., 1993, 1997). Hydroxylamine oxidoreductase purified from *Alcaligenes faecalis* strain TUD has non-heme iron and catalyzes the reduction of ferricyanide with hydroxylamine, but does not catalyze the reduction of ferricytochrome *c* with hydroxylamine (Otte et al., 1999).

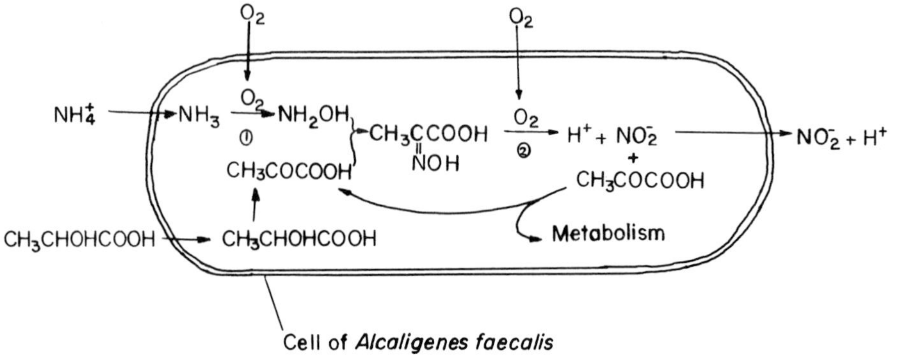


**Fig. 3.6.** A schematic presentation of nitrite formation from hydroxylamine in the cell of *Alcaligenes faecalis*. *Circled numbers*: *1*, ammonia monooxygenase; *2*, pyruvic oxime dioxygenase

*Alcaligenes faecalis* has a hydroxylamine-oxidizing enzyme which differs from the enzyme of other heterotrophic nitrifying bacteria. In this bacterium, hydroxylamine is oxidized through pyruvic oxime (Fig. 3.6). Pyruvic acid derived from lactate, succinate, and other carboxylates reacts nonenzymatically with hydroxylamine formed from ammonia to form pyruvic oxime. This compound is oxidized to nitrite and pyruvic acid by the catalysis of pyruvic oxime dioxygenase (Ono et al., 1996, 1999).

$$NH_2OH + CH_3COCOOH \xrightarrow{\text{Nonenzymatically}} \underset{\text{Pyruvic oxime}}{CH_3C(=NOH)COOH} + H_2O \quad (3.7)$$

$$CH_3C(=NOH)COOH + O_2 \xrightarrow{\text{Pyruvic oxime dioxygenase}} CH_3COCOOH + HNO_2 \quad (3.8)$$

Previously, a heterotrophic nitrifier, *Achromobacter* sp. was found to oxidize pyruvic oxime externally added to form nitrite (Quastel et al., 1952). However, as pyruvic oxime does not occur in Nature (Lang and Jagnow, 1986), the nitrification mechanism in which pyruvic oxime has to be acquired from the outside of the bacterial cells was denied. Ono et al. (1996) have shown that pyruvic oxime is formed inside the bacterial cells. Thus, an oxidation mechanism of hydroxylamine to nitrite in a heterotrophic nitrifier has been elucidated.

## 3.2 Applications of Nitrifying Bacteria

### *3.2.1 Bacterial Production of Gunpowder*

This is a story from Japanese history. Early in the Edo period (1603–1867) in Japan, the nitrifying bacteria were used to produce an ingredient of black gunpowder, niter.

Black gunpowder is a mixture of charcoal, sulfur, and niter (potassium nitrate) in a weight ratio of 13.5 : 11.9 : 74.6. Although in the Edo period much charcoal was produced and sulfur was supplied from sulfur mines, no niter was available, so niter was produced utilizing the nitrifying bacteria early in the Edo period (Aoki et al., 1978). A pit (3.6 × 3.6 × 2 m) made under the floor of a wooden house was packed with a mixture of vital organs of fish, urine of human and horse, grasses, soil, and lime. The mixture was allowed to stand for 5 years with mixing for aeration several times every summer. After 5 years of standing, the mixture was dug out leaving about 10% in the pit, and was extracted with water. For example, 1 $m^3$ of the mixture or niter soil was extracted with 500 liters of water and the extract obtained was concentrated by boiling to 15 liters. To the concentrated solution obtained was added a water extract of wood ash and the resulting mixture was further concentrated to 7.5 liters by boiling. The concentrated solution obtained was filtered through a piece of cotton cloth supported by bamboo mesh. The resulting filtrate was cooled in the outdoors during the early morning in December, and slender, yellow, wire-like crystals appeared. The crude niter crystals were twice recrystallized and colorless transparent crystals of the salt were obtained. The crystals were used as the ingredient for the black gunpowder.

### *3.2.2 Removal of Ammonia from Sewage*

To remove ammonia from drainage or sewage, ammonia-oxidizing bacteria and nitrite-oxidizing bacteria are used. The combined use of nitrifying bacteria and denitrifying bacteria is effective to remove ammonia; it produces nitrogen gas from ammonia (Sumino et al., 1992). When ammonia in the sewage is removed on an industrial scale, however, it is troublesome to switch the conditions from the aerobic treatments for nitrification to the anaerobic treatments for denitrification. Therefore, for example, polymer chips which contain both the ammonia-oxidizing bacteria and the denitrifying bacteria are used for the removal of ammonia. When the chips containing, e.g., *Nitrosomonas europaea* and *Paracoccus denitrificans* are put in the sewage and aerated, the ammonia-oxidizing bacterium gathers on the surface area of the chips while the denitrifying bacterium gathers in the inner parts of the chips, and they change ammonia to nitrogen gas in two steps (Kokufuta et al., 1988). In this case, it is unnecessary to oxidize ammonia to nitrate before the denitrification, unlike the reaction steps occurring in Nature.

If a bacterium is available which changes ammonia to nitrogen gas under aerobic conditions, such a bacterium is very convenient for the removal of ammonia in the sewage. In fact, such a bacterium, *Thiosphaera pantotropha* (now *Paracoccus pantotropha* GB17) has been found in the Netherlands (Robertson et al., 1988). However, the bacterium does not appear to be utilized actively for the removal of ammonia in the sewage. In general, the bacteria which denitrify aerobically produce usually a small amount of nitrous oxide ($N_2O$) that destroys the ozone layer of the stratosphere. This seems to be an important problem when aerobic denitrifiers are used for the bacterial removal of ammonia. Although it has been reported that *N. europaea* reduces anaerobically nitrite to nitrogen gas and nitrous oxide scarcely forms in the reactions, the bacterium has not yet been practically used to treat ammonia or nitrite (Shrestha et al., 2001). Ammonia-oxidizing bacteria have been found which anaerobically oxidize ammonia with nitrite to produce nitrogen gas (anammox bacteria) (Strous et al., 1999: Schmid et al., 2000). Although the bacteria are known to be classified in *Planctomycetes* (Tal et al., 2005), their details have not been clarified as their cultivation is very difficult. Such bacteria might be very effective in the future for anaerobic removal of ammonia from sewage. Some authors have reported attempts to use anammox bacteria for the removal of nitrogen from wastewater (Sumino et al., 2006).

## 3.3 Interaction Between Ammonia-Oxidizing and Nitrite-Oxidizing Bacteria

### *3.3.1 Was Earth Previously Polluted by Nitrite?*

An evolutionary relationship between two organisms can be recognized by comparing the amino acid sequences of the proteins which perform equivalent functions in both the organisms. For example, cytochrome *c* functions as the electron donor to cytochrome *c* oxidase in many organisms. When we compare the amino acid sequences of cytochromes *c* from various organisms, we find an evolutionary relationship among the organisms.

Cytochrome *c* molecule is composed of about 100 amino acids. When the amino acid sequence of human cytochrome *c* is compared with those of cytochromes *c* from several organisms, the number of different amino acids at the corresponding positions are 1 between human and monkey, 12 between human and horse, and 22 between human and tuna (e.g. Dickerson and Timkovich, 1975). In other words, the number of different amino acids at the corresponding positions in the sequence of cytochrome *c* is larger between the organisms which have been thought to be evolutionarily in a more remote relation mainly on the basis of morphology, or it is smaller between the organisms which have been thought to be in a more intimate evolutionary relation. Therefore, we can recognize the evolutionary relationship of the organisms by comparing the amino acid sequences of their cytochromes *c*.

The amino acid sequence of *Nitrobacter winogradskyi* cytochrome *c* [cytochrome *c*-550(s)] is more similar to those of human and horse cytochromes *c* than that of *Nitrosomonas europaea* cytochrome *c* (cytochrome *c*-552) (Table 3.4). Namely, *N. winogradskyi* seems to have appeared on Earth evolutionarily later than *N. europaea*, on the basis of the amino acid sequence of cytochrome *c* (Yamanaka and Fukumori, 1988). So it is expected that Earth might have been polluted by nitrous acid or nitrite in the period between the appearance of *N. europaea* and *N.*

**Table 3.4.** Numbers of different amino acid in the sequences of cytochromes *c* from several organisms

| Cytochrome *c* | Number of different residues | | | | | | |
|---|---|---|---|---|---|---|---|
| | (1) | (2) | (3) | (4) | (5) | (6) | (7) |
| (1) Human | 0 | | | | | | |
| (2) Rhesus monkey | 1 | 0 | | | | | |
| (3) Horse | 12 | 11 | 0 | | | | |
| (4) Tuna | 21 | 21 | 19 | 0 | | | |
| (5) Yeast (*S. cerevisiae*) | 44 | 44 | 45 | 45 | 0 | | |
| (6) *N. winogradskyi* | 62 | 61 | 59 | 61 | 72 | 0 | |
| (7) *N. europaea* | 93 | 92 | 93 | 93 | 98 | 97 | 0 |

(1) Matsubara and Smith (1963); (2) Rothfus and Smith (1965); (3) Margoliash et al. (1961); (4) Kreil (1963, 1965); (5) Narita and Titani (1969), Yaoi (1967), Lederer et al. (1972); (6) Tanaka et al. (1982); (7) Fujiwara et al. (1995)

*winogradskyi*, if *N. winogradskyi* or the nitrite-oxidizing bacteria were the only organisms that were able to remove nitrous acid or nitrite in that period, though the denitrifying bacteria might have also removed these compounds.

We find also some interesting results from the data shown in Table 3.4. Comparing the amino acid sequence of cytochrome *c* from *N. winogradskyi* with those of cytochromes *c* from human, monkey, and tuna, the numbers of the different amino acids at the corresponding positions of the sequence are found to be 62, 61, and 59, respectively. The number of different amino acids in the sequence is 72 between yeast and the bacterial cytochrome *c*. From these data, the evolutionary distance between yeast and the bacterium is farther than that between the animals and the bacterium. The number of different amino acids in the sequence of cytochrome *c* is 44 between human and yeast cytochromes *c*. By simple analysis of the figures, we are led to a conclusion that the bacterium has appeared on Earth evolutionarily after human or that yeast has appeared after human. This is a mystery of molecular evolution that needs to be solved in future studies.

### 3.3.2 An Agricultural Incident Caused by Incomplete Nitrification

Nowadays, the numbers of the ammonia- and nitrite-oxidizing bacteria present in fields, rivers, and oceans are approximately equal to one another. However, if the activity of the nitrite-oxidizing bacteria is depressed while the ammonia-oxidizing bacteria actively oxidize ammonia, the accumulation of nitrite is expected. This occurred actually in 1962 around Nangoku City, Kochi Prefecture, Japan (Nishio, 1981). Most of the vegetables in the greenhouses in this area suddenly withered. As nitrous acid of a high concentration was found in the water drops adhering to the inside of plastics sheets of the greenhouses, the scientists who inspected the agricultural incident concluded that nitrogen dioxide ($NO_2$) was responsible for the withering of the vegetables. Namely, nitrous acid of a high concentration was found to have accumulated in the soil of the greenhouses, and the pH of the soil was 5. As nitrous acid at higher concentrations is decomposed at pH less than 5.5 to nitric oxide, nitric oxide is emitted to the atmosphere in the greenhouses. Nitric oxide emitted in the atmosphere was immediately oxidized to nitrogen dioxide ($NO_2$). The resulting nitrogen dioxide caused the vegetables to wither.

Such an agricultural incident is now recognized to be caused when excessive nitrogenous fertilizers such as urea are supplied to the vegetables to stimulate their growth in the greenhouse or in an enclosed space; the ammonia-oxidizing bacteria vigorously oxidize ammonia to nitrous acid and the pH of the soil decreases to approximately 5. The nitrite-oxidizing bacteria oxidize nitrite or nitrous acid to nitrate or nitric acid while the pH of the soil is above 6, but when the pH decreases to below 6 by the active bacterial oxidation of ammonia, the nitrite-oxidizing bacteria not only stop oxidizing nitrite but also reduce nitrate to nitrite; they change into the "nitrate-reducing bacteria" at pHs below 6, as already described. So, at pHs below 6, nitrous acid accumulates and is decomposed to nitric oxide. As the causes for the incident were elucidated, the problem was solved by raising the pH of the soil to around 6 with calcium hydroxide or calcium carbonate, and by adding 2-amino-4-chloro-6-methylpyrimidine (AM, a nitrification-controlling reagent) to the soil.

### *3.3.3 Herbicides and Nitrification*

The incident described above was caused by a decrease in the pH of soil. The change of environments other than pH can also be the cause of similar incidents. Namely, the conditions that inhibit the growth of nitrite-oxidizing bacteria but do not inhibit the growth of ammonia-oxidizing bacteria may cause the accumulation of nitrite or nitrous acid in some environments. Most herbicides inhibit more or less the growth of both the ammonia-oxidizing and the nitrite-oxidizing bacteria, and their inhibitory effects disappear approximately 2 weeks after the herbicides have been added. However, paraquat (1,1′-dimethyl-4,4′-bipyridium dichloride, methyl viologen) is very special among the herbicides tested by the author; paraquat at 4 μM inhibits strongly the growth of *N. winogradskyi* but does not inhibit the growth of *N. europaea* at all. This means that when paraquat is used as a herbicide, nitrite may accumulate in the environment. However, no reports are known of nitrite accumulation by the use of paraquat as herbicide.

Nowadays, in Nature, approximately the same numbers of the ammonia-oxidizing and the nitrite-oxidizing bacteria are present in most cases, as already mentioned. So it may be attributable to the presence of approximately the same numbers of both the bacteria in Nature that nitrite does not accumulate by use of paraquat in the field. Namely, the effect of paraquat on the nitrifiers when both the ammonia-oxidizing and the nitrite-oxidizing bacteria are present at the same time may differ from that when each of the bacteria is present separately. Thus, when paraquat was added to the mix culture of *N. winogradskyi* and *N. europaea*, no nitrite accumulation was observed (Yamanaka, 1983). The result that no accumulation of nitrite occurs by use of paraquat as the herbicide in fieldwork may be attributable to a splendid "force of Nature." However, as paraquat at higher concentrations than 4 μM inhibits the growth of *N. europaea* as well as that of *N. winogradskyi*, the use of the compound as the herbicide at higher concentrations than 4 μM (approximately effective concentration in the field) may cause the pollution by ammonia.

The herbicidal effect of paraquat is attributable to the formation of superoxide anion ($O_2^-$). Superoxide anion is very toxic compound and is formed by the reaction of oxygen with paraquat radical (paraquat$^\bullet$). Plants, algae, and cyanobacteria have ferredoxin-NADP reductase to form NADPH for the reduction of carbon dioxide (see below). The chemolithoautotrophs also have NAD(P) (NAD and NADP) reductase to form NAD(P)H for the reduction of carbon dioxide. Paraquat [mid-point redox potential at pH 7.0 ($E_{m,7.0}$) = −0.43 V] radical is produced when paraquat is reduced by the catalysis of ferredoxin-NAD(P) reductase or NAD(P) reductase, which catalyzes the reduction of many compounds with $E_{m,7.0}$ of around −0.4 V. Although the aerobic organisms (and even many anaerobic organisms) have superoxide dismutase (SOD) which detoxifies superoxide anion in cooperation with catalase [ascorbate peroxidase in the case of plants (Asada, 1999)], the anion accumulates in the organisms when it is over-produced beyond the capacity of SOD.

$$2\,\text{Paraquat} + \text{NAD(P)H} \xrightarrow{\text{NAD(P) reductase}} 2\,\text{Paraquat}^\bullet + \text{NAD(P)}^+ + \text{H}^+ \quad (3.9)$$

$$\text{Paraquat}^\bullet + \text{O}_2 \xrightarrow{\text{Non-enzymatically}} \text{Paraquat} + \text{O}_2^- \quad (3.10)$$

$$2\text{O}_2^- + 2\text{H}^+ \xrightarrow{\text{SOD}} \text{O}_2 + \text{H}_2\text{O}_2 \quad (3.11)$$

$$2\text{H}_2\text{O}_2 \xrightarrow{\text{catalase*}} 2\text{H}_2\text{O} + \text{O}_2 \quad (3.12)$$

[* catalase, in general; ascorbate peroxidase, in plants]

When paraquat is sprayed on plants, superoxide anion is over-produced because paraquat produces superoxide anion, as mentioned above, and destroys plant tissues. This is the mechanism by which paraquat acts as a herbicide.

The growth of the chemolithoautotrophic bacteria can be inhibited by paraquat, because the bacteria have NAD(P) reductase. So it is rather mysterious that *N. europaea* is not affected by the compound at 4 μM. It will be attributable to hydroxylamine which is formed as an intermediate metabolite during the oxidation of ammonia by the bacterium that the inhibitory effect of paraquat is not observed with the bacterium. Hydroxylamine is a scavenger of superoxide anion (Elstner et al., 1975). Probably, *N. europaea* (and other ammonia-oxidizing bacteria) are exceptionally resistant to paraquat, but other chemolithoautotrophs all will be sensitive to paraquat. Therefore, we should take care not to disturb the soil bacteria when we use paraquat as a herbicide.

## 3.4 Reduction of Nitrate and Nitrogen Gas

### *3.4.1 Bacteria That Reduce Nitrate to Nitrogen Gas*

As mentioned above, ammonia is oxidized by the nitrifying bacteria to nitric acid via nitrous acid, and the resulting nitric acid combines with calcium carbonate in soil to produce calcium nitrate. Nitrate is a good nitrogen source for plants. However, in the soil, the denitrifying bacteria are present which reduce nitrate to nitrogen gas under oxygen-deficient conditions and compete for nitrate with plants, because the bacteria oxidize organic compounds with the salt to form ATP. As the denitrifying bacteria reduce nitrate very actively especially in the Torrid Zone, nitrogenous fertilizers used are very quickly changed finally to nitrogen gas and the effect of the fertilizers is hardly observed there.

Most of the denitrifying bacteria oxidize organic compounds with nitrate under oxygen-deficient conditions to form ATP; i.e., they are chemoheterotrophs. Among the denitrifiers, chemolithoautotrophs are also found; they oxidize inorganic compounds like thiosulfate (Aminuddin and Nicholas, 1973) (see p.72), ferrous iron (Hafenbradl et al., 1996) (see p. 91), and hydrogen gas (Suzuki et al., 2001) with nitrate. Under oxygen-deficient conditions, nitrate is inevitable for the denitrifying bacteria to form ATP. Therefore, the bacteria need to utilize nitrate before plants utilize it. In the bacterial oxidation of organic compounds with nitrate, the salt is reduced to nitrogen gas ($N_2$) via nitrite ($NO_2^-$), nitric oxide (NO), and nitrous oxide (dinitrogen monoxide) ($N_2O$), and each of the intermediate metabolites also oxidizes organic compounds. When the organic compounds are oxidized by nitrate and by the intermediate nitrogen compounds formed, energy is liberated which is available for biosynthesis of ATP (Fig. 3.7). The denitrification named on the basis of only the change of nitrogen compounds is now called nitrate respiration consid-

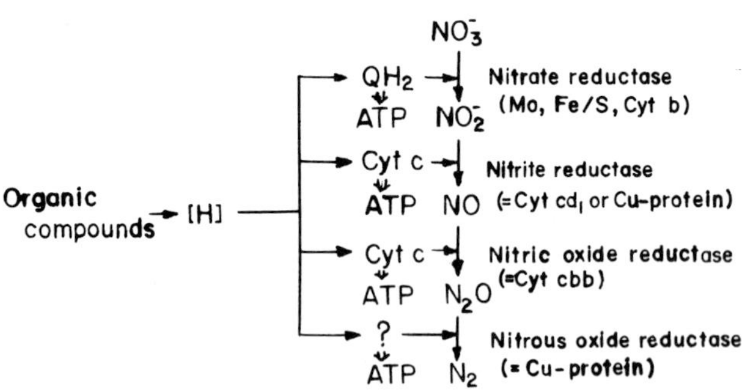


**Fig. 3.7.** A schematic presentation of the reaction processes in the nitrate respiration. *[H]*, hydrogen atom bound to certain compounds; *Q*, ubiquinone; *Cyt*, cytochrome. *Downward arrow* means that ATP is biosynthesized using the energy released by the corresponding electron transfer systems. Nitric oxide reductase is also known to use quinol as the electron donor (see text and Chapter 2, p. 13)

ering the biosynthesis of ATP coupled to the reduction of nitrate and the intermediate nitrogen compounds formed during the nitrate reduction.

In the heterotrophic nitrate respiration, in most cases hydrogen atoms derived from organic compounds ([H], mostly in the form of NADH in the case of heterotrophic denitrifiers) are first oxidized with nitrate. The reaction is catalyzed by nitrate reductase. The enzyme contains Mo, [$Fe_4S_4$] and [$Fe_2S_2$] clusters (Fe/S), and cytochrome *b* (Chaudhry and MacGregor, 1983). Mo is present in the enzyme as molybdenum cofactors combining with molybdopterin or molybdopterin guanine dinucleotide. The enzyme catalyzes the reduction of nitrate to nitrite with ubiquinol or menaquinol ($QH_2$) as the electron donor.

Next, nitrite is reduced to nitric oxide by the catalysis of nitrite reductase. Two kinds of nitrite reductases are known in the denitrifying bacteria ($N_2$ forming); cytochrome $cd_1$-type enzyme (Yamanaka et al., 1960, 1961, 1963; Yamanaka and Okunuki, 1963a,b,c; Yamanaka, 1964) and copper protein-type enzyme (Iwasaki and Matsubara, 1972). However, no case has been found in which one species of the denitrifying bacterium has both types of the enzymes simultaneously (Coyne et al., 1989).

In 1939, Yamagata indicated that an enzyme which reduced nitrite was present in the cell-free extract of *Bacillus pyocyaneus* (now *Pseudomonas aeruginosa*), but the finding was suspected by many researchers worldwide. The author and his coworkers (Yamanaka et al., 1960; Yamanaka, 1964) have demonstrated that his finding was true on the basis of the purification and characterization of the enzyme. Cytochrome $cd_1$ of *Pseudomonas aeruginosa* catalyzes the reduction of nitrite to nitric oxide with ferrocytochrome *c*-551 and reduced azurin of the bacterium. The cytochrome has one molecule each of heme C and heme $D_1$ in the molecule with molecular mass of 67 kDa (Nagata et al., 1970) (60 kDa from DNA, Silvestrini et al., 1989). As the cytochrome usually occurs as dimer in the solution (Kuronen and Ellfolk, 1972), its monomer was previously thought not to have the enzymatic activity. However, the monomeric cytochrome formed by a chemical modification still shows activity (Silvestrini et al., 1995), and it has now been verified that the monomeric cytochrome has such activity.

Cytochrome $cd_1$ is the first cytochrome that has been found to have two kinds of hemes, hemes C and $D_1$ in one molecule. After the discovery of cytochrome $cd_1$, many cytochromes have been found which possess two or three kinds of hemes in the molecule; cytochromes *bd*, $ba_3$, $baa_3$, *bo* (or $bo_3$), $caa_3$, $cbb_3$, and *aco* (or *cao* or $cao_3$) (see Yamanaka 1992).

The copper-protein type nitrite reductases have been purified from several denitrifying bacteria. They have 2–4 copper atoms in the molecule; one copper per subunit (30–40 kDa). The enzymes catalyze the reduction of nitrite to nitric oxide, but the physiological electron donors for the enzymes have not been clarified, though they are said to be cytochrome *c* in some cases and a copper-protein in some cases (see Otsuka and Yamanaka, 1988).

Nitric oxide formed from nitrite by the catalysis of nitrite reductase is next reduced to nitrous oxide ($N_2O$). The reduction of nitric oxide is catalyzed by nitric oxide reductase (NO reductase) which is a cytochrome *cbb* (Carr and Ferguson,

1990; Fujiwara and Fukumori, 1996). Nitric oxide reductase is also known, which lacks heme C and uses quinol as the electron donor (Suharti et al., 2001; de Vries et al., 2003). The cytochrome *cbb*-type enzyme has a molecular structure similar to the structure of the CuB binding portion in cytochrome *c* oxidase (Saraste and Castresana, 1994; Van der Oost et al., 1994; Zumft et al., 1994). Moreover, quinol NO reductase from *Bacillus azotoformans* is known to contain CuA. Nitrous oxide is further reduced to nitrogen gas ($N_2$) by the catalysis of nitrous oxide reductase ($N_2O$ reductase) which is a multi-copper protein (Zumft and Matsubara, 1982). The structure of the copper-binding portion in the enzyme has been reported also to be similar to the structure of the CuA binding portion of cytochrome *c* oxidase (Charnock et al., 2000).

## 3.4.2 Nitric Oxide Is also Produced in Human Tissues

Until the end of twentieth century, it was thought that nitric oxide is formed only by the denitrifying bacteria (and by some plants to a lesser extent). However, recently nitric oxide has been found to be formed in the human tissues and to act as various physiological effectors (e.g., Vincent, 1995). For example, it expands the coronary artery of heart. It has been known for a long time that nitroglycerin is effective in curing an attack of angina pectoris. How was it discovered that nitroglycerin cures attacks of angina pectoris? Its curative effect was reported to have been found from the results that mine workers felt healthy conditions of the heart on days when they blasted ores with dynamite. Although the author does not know whether or not this story is true, he has heard that in actual fact the employees in the factory producing dynamite complained of pain in their heart every Monday morning. In other words, as they inhale daily the vapor of nitroglycerin the blood walls of their coronary arteries are always in an expanded state. On Sunday, as they do not inhale the vapor, their arteries are in a shrunken state and they feel the attack of angina pectoris on Monday morning. However, it was not known for a long time why nitroglycerin cures angina pectoris attacks. It is now known that nitric oxide is formed from nitroglycerin by the catalysis of cytochrome P-450 (Servent et al., 1989).

By what mechanism is nitric oxide formed in the human tissues? It has been known for a long time that the amount of nitrate excreted by the human body mainly as urine is larger than that of the salt taken in as foods. The cause of the difference was an issue of debate. Some researchers thought that the nitrifying bacteria in the intestine oxidized ammonia derived from amino acids. However, others argued against the idea, insisting that the nitrification by the bacteria could not occur in the intestine as the partial pressure of oxygen is considerably low there. Actually, it is now known that nitrate is reduced in the intestine (Saul et al., 1981).

While the causes for the nitrification in the human body have not been elucidated, germfree mice were exploited. The amount of nitrate excreted was larger than that of the salt taken in as the foods also in germ-free mice. Namely, it has become plausible that nitrate is produced in animal tissues. Meanwhile, it has been found that nitric oxide is produced from L-arginine by the catalysis of nitric oxide synthase (NOS) in the human tissues. Nitric oxide disappears by oxidation after its physiological functions. One of the oxidation processes of the compound is the reaction with oxyhemoglobin [Hb($Fe^{2+}$)—$O_2$]; nitrate and methemoglobin [MetHb($Fe^{3+}$)] are formed as the result.

$$NO + Hb(Fe^{2+})\text{—}O_2 \rightarrow NO_3^- + MetHb(Fe^{3+}) \quad (3.13)$$

It is explainable by the reaction that nitrate is formed in the human tissues, though other reactions will occur which form nitrate from nitric oxide.

Nitric oxide is rapidly oxidized to nitrogen dioxide ($NO_2$) by molecular oxygen in air, as already mentioned above. So it seems that nitric oxide once formed is too rapidly oxidized to nitrogen dioxide to function, as the concentration of molecular oxygen in the human tissues is around 200 μM. However, as the concentration of nitric oxide physiologically formed in the tissues is very low (ca. 0.1 μM), the rate of the oxidation of nitric oxide to nitrogen dioxide is extremely low even if the concentration of molecular oxygen is approximately 200 μM, and nitric oxide can remain for approximately 80 min on calculation. Actually, nitric oxide remains only for 6–8 s in the tissues, as nitric oxide is probably consumed by several processes. The period of 6–8 s for the stay of nitric oxide in the tissues seems reasonable, as nitric oxide may be harmful for the tissues if it remains for longer periods than this.

In tissues, nitric oxide activates guanyl cyclase which catalyzes the formation of cyclic guanosine 3′,5′-monophosphate (cGMP) from guanosine 5′-triphosphate (GTP). Cyclic GMP functions as a trigger to activate various enzymes and proteins, and causes, for example, the relaxation of smooth muscles. When the smooth muscles composing the wall of arteries are relaxed, the coronary arteries are expanded and the blood in the arteries flow smoothly. This cures the angina pectoris. When the smooth muscles composing the arterial walls and sponge tissues of the penis relax, the penis becomes erect. When cGMP phosphodiesterase hydrolyzes cGMP, the relaxation of the smooth muscles finishes. Viagra inhibits phosphodiesterase and makes the erection of the penis continue for longer periods (Fig. 3.8).

### *3.4.3 Bacteria Reducing Nitrogen Gas to Ammonia*

If nitrate formed by the oxidation of ammonia is completely changed to nitrogen gas by the denitrifying bacteria, plants cannot utilize nitrogen sources and the surface of Earth might become sterile soil and bare mountains. However, actually nitrogen-fixing bacteria reside which reduce nitrogen gas to ammonia, so plants can grow utilizing the ammonia and its related nitrogen compounds. This ammonia does not mean ammonia itself or ammonia compounds contained in the chemical fertilizers; ammonia (or ammonium ion) formed by the bacteria is immediately

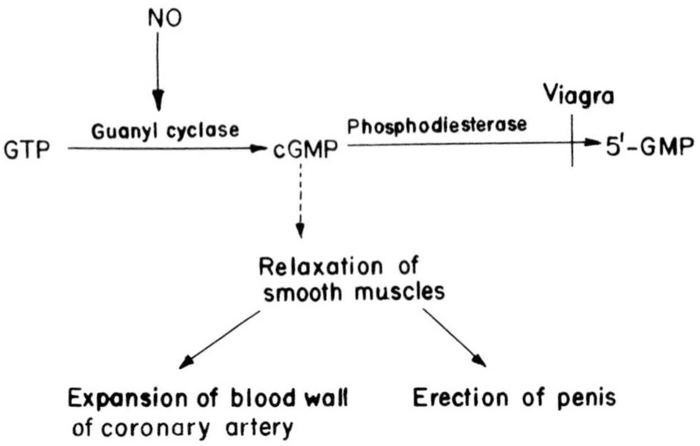


**Fig. 3.8.** A schematic presentation of the effect of nitric oxide on the relaxation of smooth muscles. *GTP*, guanosine 5′-triphosphate; *5′-GMP*, guanosine 5′-monophosphate; *cGMP*, cyclic guanosine 3′,5′-monophosphate

incorporated into organic compounds, mainly into amino acids at once when it is formed.

The reduction of nitrogen gas to ammonia is catalyzed by nitrogenase or more exactly by nitrogenase complex. Nitrogenase complex is composed of a reductase component (Fe protein) and a nitrogenase component (MoFe protein). The reductase component is a dimer of two identical subunits with the molecular mass of around 30 kDa. The dimer coordinate with a single [$Fe_4S_4$] cluster (Georgiadis et al., 1992). The nitrogenase component is composed of two kinds of subunits ($\alpha$ and $\beta$) with approximately the same molecular mass (around 60 kDa). Each of the $\alpha$ subunits has a MoFe-cofactor which is the nitrogen reducing site. The factor is composed of [$Fe_3MoS_3$] and [$Fe_4S_3$] linked through three inorganic sulfur atoms (Chan et al., 1993; Karlin, 1993). Each of the $\beta$ subunits has two [$Fe_4S_4$] clusters linked through two sulfur atoms of cysteine residues (P-cluster), and the functional nitrogenase component has the subunit structure of $\alpha_2\beta_2$ (Kim and Rees, 1992). Electrons from the [$Fe_4S_4$] cluster in the reductase component are transferred to the P-cluster in the $\beta$ subunit of the nitrogenase component and then to the MoFe-cofactor where nitrogen is reduced. Although molybdenum in the enzyme plays an important role in the reduction of nitrogen, the metal is replaced by vanadium when the nitrogen-fixing bacterium *Azotobacter vinelandii* is grown under molybdenum-deficient conditions (Eady, 1988).

Twelve molecules of ATP are theoretically necessary for the reduction of one molecule of nitrogen gas catalyzed by nitrogenase. But actually at least 16 molecules of ATP are required for the enzymatic reduction of nitrogen gas, because hydrogen gas is inevitably produced during the reduction of nitrogen gas and ATP is necessary also for the production of hydrogen gas.

$$N_2 + 8e + 16ATP + 16H_2O \xrightarrow{\text{Nitrogenase}} 2NH_3 + H_2 + 16ADP + 16P_i + 8H^+ \quad (3.14)$$

For nitrogen fixation, many ATP molecules are consumed as mentioned above. As ATP is the source of the energy necessary for the life processes, nitrogen-fixing bacteria stop fixing nitrogen to save energy when nitrogen compounds such as ammonia and amino acids are available.

The nitrogen-fixing bacteria, in general, seem to be evolutionarily older than other bacteria that do not fix nitrogen. Therefore, one question arises: if the primordial organisms that lived in the environments where various nitrogen-containing compounds such as amino acids were present as the results of chemical evolution, they might not have needed the nitrogen fixation, i.e., they seem not to have needed nitrogenase. However, nitrogenase has not only the catalytic activity to reduce nitrogen to ammonia but also the activity to reduce hydrogen cyanide (HCN) to ammonia and methane. Since much hydrogen cyanide is supposed to have occurred in the atmosphere of the primordial Earth, nitrogenase might have first been used for the detoxification of hydrogen cyanide (Silver and Postgate, 1973), and then it would have been used as the catalyzer for the reduction of nitrogen after the nitrogen-containing organic compounds were exhausted, because the enzyme has had the structural ability to reduce nitrogen. In any case, at present, it is true that the nitrogen-fixing bacteria fix nitrogen to supplement nitrogen compounds which are lost by denitrification.

Nitrogenase reduces acetylene as well as nitrogen and hydrogen cyanide. Previously, because the assay of ammonia was not easy to carry out, studies on nitrogen fixation were developed slowly. After the finding that nitrogenase also reduces acetylene, the studies on the nitrogen fixation developed very quickly, because the nitrogen-fixing activity of nitrogenase is determined by measuring ethylene formed from acetylene with gas chromatography. However, as ethylene is usually not further metabolized while ammonia is further metabolized in the organisms, we should pay attention to the difference between the ethylene and ammonia assays of nitrogen-fixing activity in vivo.

As nitrogenase is very labile in the presence of oxygen, its purification should be performed anaerobically. The purification of the enzyme was rapidly developed after the anoxic box became available. Among the nitrogen-fixing bacteria there are strict aerobic bacteria. It is of interest to know how the aerobic bacteria can fix nitrogen because nitrogenase is labile in the presence of oxygen and the aerobic bacteria need oxygen for respiration.

#### (a) Rhizobia

The roots of the leguminous plants have many nodules in which the nitrogen-fixing bacteria, rhizobia, are present. Although the rhizobia reside also in soil in a free-living state, they do not fix nitrogen in this state. The bacteria fix nitrogen only in the nodules (symbiotic state). Why do they fix nitrogen only when they are in the symbiotic state? It was not until 1975 that the reason was recognized. In the root nodules, leghemoglobin occurs, which combines oxygen very strongly (Wittenberg et al., 1972; Wittenberg, 1974). Thus, the insides of the nodules are kept anaerobic and nitrogenase in the rhizobia function regardless of the effect of oxygen, while the bacteria can breathe tearing off oxygen from oxyleghemoglobin (leghemoglobin bound to oxygen). Namely, the root nodules are the place where anaerobiosis and aerobiosis coexist.

Leghemoglobin is composed of heme and globin, a protein similar to animal hemoglobin. However, the affinity for oxygen of leghemoglobin is much higher than that of animal hemoglobin (Wittenberg et al., 1972; Wittenberg, 1974). The function of animal hemoglobin is to combine and to release oxygen; i.e., hemoglobin is a carrier of oxygen, while the function of leghemoglobin is only to combine oxygen. The globin part of leghemoglobin is biosynthesized by leguminous plants (Sidloi-Lumbroso and Schulman, 1977; Baulcombe and Verma, 1978), while for heme this is done by rhizobia (Cutting and Schulman, 1969; O'Brian et al., 1987).

Although leghemoglobin was first found by Kubo in 1939 in Japan, the real studies on leghemoglobin were started mainly by foreign researchers around 1970 (Wittenberg et al., 1972; Appleby et al., 1973). The amino acid sequences of leghemoglobins were found to be similar to those of animal hemoglobins (Ellfolk and Sievers, 1974). Furthermore, the amino acid sequence of a bacterial hemoglobin (hemoglobin of *Vitreoscilla* spp.) (Orii and Webster, 1986) has been found by the cooperation of Japanese and American researchers to have a sequence similar to those of animal hemoglobins (Wakabayashi et al., 1986). Therefore, it is expected that hemoglobin had occurred before the evolutionary branching of bacteria, plants, and animals.

### (b) Azotobacter

The bacteria of *Azotobacter* genus are also well known as nitrogen fixers. The bacteria fix nitrogen in soil in the presence of oxygen at the atmospheric partial pressure, unlike rhizobia which fix nitrogen only in a shelter of nodules. The reason why *Azotobacter vinelandii* can fix nitrogen in aerobic soil is that nitrogenase of the bacterium is protected by Shethna protein (Shethna et al., 1968; Brill, 1980). It is also known that, in the cells of the bacterium, the level of NADH: ubiquinone oxidoreductase not coupled to ATP biosynthesis becomes higher under higher concentrations of molecular oxygen (Bertsova et al., 2001). This is also a device to diminish oxygen concentration in the microenvironment around nitrogenase. Moreover, cytochrome *bd* is known to reduce oxygen concentration around nitrogenase (Kelly et al., 1990; Poole and Hill, 1997). Although the bacteria of *Azotobacter* genus have devised the method to fix nitrogen under oxygen at the atmospheric partial pressure, most of the aerobic free-living nitrogen fixers generally fix nitrogen only in the environments where the partial pressure of oxygen is relatively reduced. For example, as the partial pressure of oxygen is fairly low in soil of rice paddies, the bacteria of *Azospirillum* genus fix nitrogen around the roots (rhizosphere) of rice plants. As the roots of rice plants excrete a substance which attracts the bacteria, the bacteria gather around the roots. When the bacteria die and are decomposed, they supply nitrogen compounds to the rice plants. Such a relation as seen between the bacteria of *Azospirillum* genus and rice plants is called a loose symbiosis. Some of the bacteria of *Azospirillum* genus seem to fix nitrogen in the rhizospheres of wheat, barley, corn, and sugarcane, though the partial oxygen pressure in the rhizospheres is not so low as that in the rice paddies. Thus, it is reported that the crops of corn, barley, and wheat are more abundant when the bacteria of *Azospirillum* genus are applied to the fields where these plants grow (Bashan et al., 2004). It has been recently found that nitrogen-fixing bacteria reside in the tissues of plants like sugarcane and sweet potato and supply nitrogen to the plants (Cavalcante and Dobereiner, 1988; Paula et al., 1991; Reis et al., 2000).

### (c) Cyanobacteria

Organisms which have devised another aerobic nitrogen-fixing ways are cyanobacteria. Cyanobacteria belong to bacteria on the basis of the cellular structures, while they have the photosynthetic mechanisms which are similar to those of algae and plants. Namely, they evolve oxygen during photosynthesis. Many of them fix nitrogen and they have to protect their nitrogenase from oxygen produced during the photosynthesis. Filamentous cyanobacteria have special cells called heterocysts. These cells have fairly thicker cell walls as compared to the ordinary cells, vegetative cells. When the cyanobacteria are observed under a fluorescent microscope, the ordinary cells emit red fluorescence while the heterocysts do not emit the fluorescence because they do not have a pigment protein, phycocyanin. The heterocysts do not evolve oxygen during photosynthesis and nitrogenase is included exclusively in the cells (e.g., Fay, 1992), i.e., the heterocysts are special cells for the nitrogen fixation (Fay et al., 1968; Peterson and Wolk, 1978; Houchins and Hind, 1983). However, many cyanobacteria are also found which do not seem to have heterocysts. The devices in nitrogen fixing for nonheterocystous cyanobacteria have not been clarified.

# Chapter 4
# Sulfur Circulation on Earth and Bacteria

Sulfur (S) on Earth changes its form e.g. from hydrogen sulfide ($H_2S$) to sulfate via elemental sulfur (S°), and sulfate is changed again to hydrogen sulfide as shown in Fig. 4.1. Furthermore, hydrogen sulfide forms pyrite ($FeS_2$) and sulfuric acid is formed from this compound. Most of these processes are performed by the actions of bacteria. Hydrogen sulfide is also evolved from hot springs and volcanoes, and occurs when dead animals, the excreta of animals, and dead plants are decomposed by bacteria. The compound is oxidized to sulfuric acid by the sulfur-oxidizing bacteria and photosynthetic sulfur bacteria via elemental sulfur. The change of hydrogen sulfide to elemental sulfur occurs also abiotically in the presence of molecular oxygen. When the oxidation of hydrogen sulfide to sulfuric acid occurs in soil, the sulfuric acid formed usually reacts with calcium carbonate to produce calcium sulfate. Calcium sulfate serves as the sulfur source for plants, while it is also reduced to hydrogen sulfide by the sulfate-reducing bacteria. When the sulfuric acid formed is not neutralized and remains as an acid, it pollutes the environment acidically and sometimes corrodes concrete. When elemental sulfur formed during the oxidation of hydrogen sulfide is not further oxidized, it produces sulfur ores or mountains of elemental sulfur.

Moreover, dimethyl sulfide [$(CH_3)_2S$] should not be ignored in the circulation of sulfur on Earth. Sulfur of maximally about $80 \times 10^{12}$ g, $70 \times 10^{12}$ g, and $45 \times 10^{12}$ g is evolved per year, at present, from the surface of Earth as sulfur dioxide, hydrogen sulfide, and dimethyl sulfide, respectively (Uchida, 1995). The greater part of dimethyl sulfide is produced by marine algae and marine cyanobacteria. When these organisms are smashed by grinding with, e.g., sea sands, they smell of the beach. The smell is attributable to dimethyl sulfide evolved from the organisms. Even cyanobacteria grown inland smell of the beach when they are smashed by grinding. Dimethyl sulfide is degraded by bacteria such as *Thiobacillus* and *Hyphomicrobium*, and finally sulfuric acid is formed (Visscher and Taylor, 1993).

## 4.1 Bacteria Forming Hydrogen Sulfide

As mentioned above, hydrogen sulfide is evolved from within the Earth, such as from hot springs and volcanoes, and is derived from the decomposition by bacteria


doi: 10.1007/978-4-431-78541-5_4, 

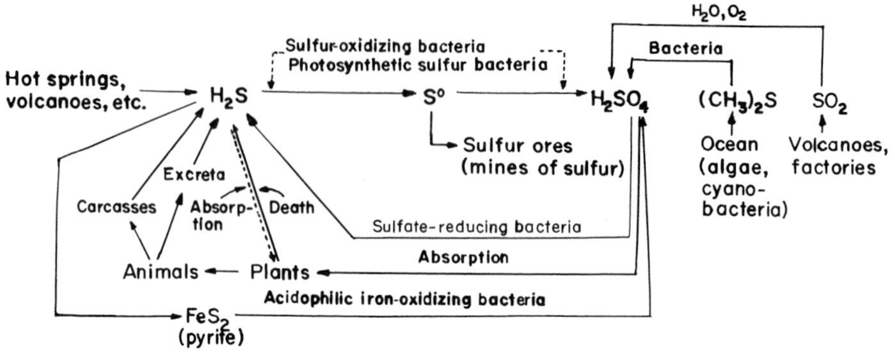


**Fig. 4.1.** A summary of natural circulation of sulfur. As the intake of hydrogen sulfide by plants is limited to non-photosynthetic plants (fungi, bacteria, etc.), the *arrow* indicating the sulfur transport from hydrogen sulfide to plants is shown by a *dotted line*

of dead animals, the excreta of animals, and dead plants. Moreover, hydrogen sulfide is produced from the reduction of sulfate by sulfate-reducing bacteria. The bacterial formation of hydrogen sulfide brings about environmental pollution; the gas itself causes a bad smell and the corrosion of some metals, and sulfuric acid formed from its oxidation by the sulfur-oxidizing bacteria causes acidic pollution and corrosion of concrete.

Around fouled rivers it sometimes smells of hydrogen sulfide because the sulfate-reducing bacteria reside on the bottom of the river and oxidize organic compounds with sulfate, resulting in the formation of hydrogen sulfide. The sulfate-reducing bacteria are strict anaerobes and form adenosine triphosphate (ATP) by oxidizing organic compounds with sulfate in place of oxygen. This process is called sulfate respiration, as already mentioned in the preceding chapter.

### *4.1.1 Bacterial Reduction Mechanisms of Sulfate*

Most of the sulfate-reducing bacteria anaerobically oxidize organic compounds with sulfate ion to obtain energy for the life processes, i.e., they are chemoheterotrophic bacteria. However, some groups of the sulfate-reducing bacteria grow by oxidizing hydrogen with a sulfate ion, i.e., they are chemolithoautotrophic bacteria (Fauque et al., 1991). Chemoheterotrophic sulfate-reducing bacteria usually oxidize lactate anaerobically with sulfate and form ATP through the substrate level phosphorylation, i.e., ATP is formed from acetyl phosphate ($CH_3COOPO_3H_2$). Previously, the bacteria were thought to grow by fermentation. However, as shown in Fig. 4.2, it has been demonstrated with *Desulfovibrio desulfuricans* that ATP formed by the substrate level phosphorylation is exhausted by the activation of sulfate to adenylylsulfate (APS) and subsequent phosphorylation of adenosine monophos-

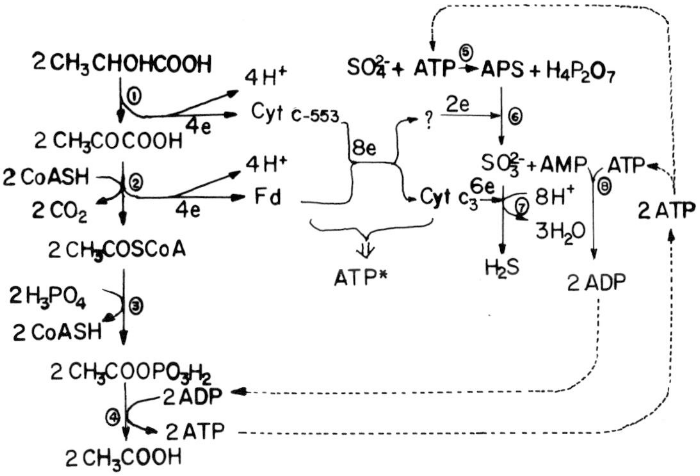


**Fig. 4.2.** The oxidation mechanisms of lactate by sulfate in the sulfate-reducing bacteria of *Desulfovibrio* genus. *Circled numbers*: *1*, lactate dehydrogenase (cytochrome *c*-553); *2*, pyruvate-ferredoxin 2-oxidoreductase (CoA-acetylating); *3*, phosphate acetyltransferase; *4*, acetate kinase; *5*, sulfate adenylyltransferase; *6*, adenylylsulfate reductase; *7*, sulfite reductase; *8*, adenylate kinase. **ATP*: adenosine 5′-triphosphate is also biosynthesized by the catalysis of ATP synthase using the energy liberated by the electron transfer around this part

phate (AMP) formed by the reduction of APS. Therefore, researchers have come to consider that the phosphorylation other than at the substrate level should occur (Vosjan, 1970) in the sulfate-reducing bacteria. Thus, it has been found that ATP is formed coupled also with an electron transfer, as will be mentioned below. However, as the bacteria of *Desulfotomaculum* genus have acetate kinase (diphosphate), they can grow only on the substrate level phosphorylation (Liu and Peck, 1981a).

As mentioned above, some of the sulfate-reducing bacteria can grow by oxidizing hydrogen gas with sulfate (Badziong and Thauer, 1978; Badziong et al., 1978); it is expected they get energy for their life processes by way of phosphorylation rather than substrate-level phosphorylation because they cannot utilize acetyl phosphate in the formation of ATP during the oxidation of hydrogen with sulfate. The bacteria are expected to produce ATP coupling with the electron transfer. Thus, Peck (1966) demonstrated that ATP was formed by the oxidation of hydrogen with sulfite using the membrane fractions of *Desulfovibrio gigas*. Therefore, the oxidation of hydrogen and organic compounds with sulfate in the sulfate-reducing bacteria, i.e., the reduction of sulfate by the bacteria, is called sulfate respiration.

In Fig. 4.2, the oxidation mechanisms of lactate with sulfate in the bacteria of *Desulfovibrio* genus is shown. In the reduction of sulfate, first the compound is changed into adenylylsulfate (APS) using ATP. APS is reduced by the catalysis of adenylylsulfate reductase to sulfite and AMP. The sulfite formed is reduced to hydrogen sulfide by the catalysis of sulfite reductase. As shown in Fig. 4.2, 2 molecules of lactate are oxidized to 2 molecules of acetate coupled to the reduction of 1 molecule of sulfate, and 2 molecules of ATP are formed through acetyl phosphates. One of the 2 molecules of ATP is used to form adenylylsulfate and the other one molecule is used to change AMP to ADP which accepts the phosphate group from acetyl phosphate. Although the oxidation of 2 molecules of lactate and the reduction of 1 molecule of sulfate balance each other as far as electrons, the bacteria cannot obtain ATP for their life processes as ATP formed is used up by the processes for the reduction of sulfate. The ATP necessary for life processes of the bacteria is formed coupled with the electron transfer from cytochrome *c*-553 and ferredoxin to APS through an as yet unidentified electron carrier and to sulfite through cytochrome $c_3$. The electron donor for sulfite reductase is cytochrome $c_3$ (Steuber et al., 1994), while that for APS reductase has not yet been identified. Sulfite reductase catalyzes the reduction of sulfite to hydrogen sulfide without intermediates when sufficient electrons are supplied. When the bacteria grow on hydrogen and sulfate, ATP is biosynthesized using the energy liberated by the electron transfer, at least, between hydrogenase and sulfite reductase, considering that ATP is formed by the oxidation of hydrogen gas with sulfite in the membrane fractions of *D. gigas*.

### *4.1.2 Components Participating in Bacterial Reduction of Sulfate*

#### (a) Hydrogenases

Three kinds of hydrogenases are known to be present in the sulfate-reducing bacteria: [Fe] hydrogenase, [Fe,Ni] hydrogenase and [Fe,Ni,Se] hydrogenase. Each of them uses cytochrome $c_3$ as the electron donor and acceptor. [Fe] and [Fe,Ni] enzymes show stronger $H_2$ evolving activity than $H_2$ absorbing activity, while [Fe,Ni,Se] enzyme shows stronger $H_2$ absorbing activity than $H_2$ evolving activity. Among the bacteria of *Desulfovibrio* genus, some have one of the three kinds of enzymes, some have two kinds, and some have all of the three kinds (Fauque et al., 1991).

#### (b) Cytochromes

Cytochrome $c_3$ is characteristic of the sulfate-reducing bacteria. The cytochrome has four heme C molecules in the molecule (13 kDa) (Yagi and Maruyama, 1971). Both the 5th and 6th axial ligands of the four heme C molecules in the cytochrome are histidine residues at pH 6.0(Higuchi et al., 1981). The midpoint redox potentials at pH 7.0 ($E_{m,7.0}$) of the four heme molecules vary with the heme ranging from −0.220 to −0.390 V (Sokol et al., 1980; Benosman et al., 1989). Two of the four heme molecules are bound to two cysteine residues (Cys) through a thioether link in the sequences of -Cys-A-B-Cys-, while the remaining two heme molecules are bound to two cysteine residues in the sequences of -Cys-A-B-C-D-Cys$^-$ (Ambler et al., 1969; Higuchi et al., 1981, 1984; Czjzek et al., 1994). The axial ligands to heme C molecules vary with pH; at pH 7–8, the 6th ligands of two heme molecules are changed to lysine residues ($\varepsilon$-$NH_2$ of lysine residue) and at pH 9, the 6th ligands of all of four heme molecules are changed to lysine residues (Kitagawa et al., 1977).

It was shown in 1968 (Yagi et al., 1968) that cytochrome $c_3$ acts as the electron donor for hydrogenase. However, it was not until 1994 that the cytochrome was found to function perfectly as the electron donor for sulfite reductase. Steuber et al. (1994) has demonstrated that sulfite reductase solubilized with the aid of a detergent from *D. desulfuricans* catalyzes the reduction of sulfite to hydrogen sulfide without any intermediates using reduced cytochrome $c_3$ as the electron donor.

In the sulfate-reducing bacteria there is another cytochrome which resembles cytochrome $c_3$ in the spectral properties and redox potential but differs from this cytochrome in molecular mass; this is cytochrome $c_3$ (26 kDa) which has eight heme C molecules in the molecule. The cytochrome molecule is composed of two polypeptides of 13 kDa (Loufti et al., 1989). On the basis of the amino acid sequence, however, the 13 kDa polypeptide differs from cytochrome $c_3$ (Guerlesquin et al., 1982; LeGall and Peck, 1987; Loufti et al., 1989). *Desulfovibrio gigas* cytochrome $c_3$ (26 kDa) molecule is composed of two 13 kDa molecules bound to each other by an S–S bond (Bruschi et al., 1996). It is claimed that cytochrome $c_3$ (26 kDa) is very effective as the electron donor for thiosulfate reductase (Hatchikian et al., 1972).

In the bacteria of *Desulfovibrio* genus, cytochrome *c*-553 functions as the electron acceptor for lactate dehydrogenase (Ogata et al., 1981). The cytochrome is a monoheme cytochrome *c* of 9 kDa and seems to belong to cytochrome $c_6$ (Moore and Pettigrew, 1990; Yamanaka et al., 1995). When the stereo structure of cytochrome *c*-553 is compared to that of cytochrome $c_6$, the regions of the N-terminal and C-terminal, the upper parts of the molecules, resemble each other, while the lower parts of the molecules are considerably different from each other (Nakagawa et al., 1990). Cytochrome *c*-553 is known to function also as the electron acceptor for formate dehydrogenase (Yagi, 1979) in *Desulfovibrio vulgaris* Miyazaki. Incidentally, the electron acceptor for pyruvate synthase [pyruvate-ferredoxin 2-oxidoreductase (CoA-acetylating)] is ferredoxin in this bacterium (Ogata et al., 1988).

In the sulfate-reducing bacteria, hexaheme cytochrome *c* is present in addition to the cytochromes mentioned above. The cytochrome *c* is a nitrite reductase which catalyzes the reduction of nitrite to ammonium ion (Liu and Peck, 1981b). A cytochrome *c* of 75 kDa having 16 heme C molecules is obtained from *D. vulgaris* (Higuchi et al., 1987) and *D. gigas* (Chan et al., 1994). Furthermore, a cytochrome *c* having 12 heme C molecules is also obtained from *D. desulfuricans* (Liu et al., 1988). However, the functions of these multiheme C-containing cytochromes have not yet been clarified.

### (c) Adenylylsulfate Reductase

In the reduction of sulfate by the sulfate-reducing bacteria, sulfate is first changed to adenylylsulfate (adenosine-5′-phosphosulfate, APS) by the catalysis of sulfate adenylyltransferase and then reduced to sulfite by the catalysis of adenylylsulfate reductase (Ishimoto, 1959; Peck, 1959). Adenylylsulfate reductase isolated from *D. vulgaris* has molecular mass of 220 kDa in the monomeric state and has one molecule of FAD, and 12 atoms of each of nonheme iron and inorganic sulfide (Bramblett and Peck, 1975). The nonheme iron and inorganic sulfide seems to construct $[Fe_4S_4]$ cluster (Lampreia et al., 1990). Although the enzyme catalyzes the reduction of adenylylsulfate to sulfite with methyl viologen radical in vitro, it has not yet been demonstrated whether or not cytochrome $c_3$ or ferredoxin functions as the electron donor for the enzyme.

### (d) Sulfite Reductase

Although many sulfate-reducing bacteria were known for a long time to have a protein pigment grayish green in color, desulfoviridin (Postgate, 1956), its function was demonstrated only in 1971 to be sulfite reductase (Lee and Peck, 1971; Kobayashi et al., 1972). However, there remained many mysterious properties about the enzyme even after its function was uncovered. Namely, the enzyme catalyzed the reduction of hydrogensulfite ion (or bisulfite ion) only with viologen radical as the electron donor, and in many cases, hydrogensulfite was enzymatically reduced to hydrogen sulfide in vitro via trithionate and thiosulfate though sometimes the compound was reduced directly to hydrogen sulfide without using the intermediates. The absorption spectrum of the purified enzyme was not changed on addition of reducing or oxidizing reagents although it changed in vivo on oxidation and reduction. Furthermore, when the enzyme was treated with HCl-acetone, porphyrin but not heme was extracted. However, finally, Murphy and Siegel (1973) demonstrated that dissimilatory sulfite reductase of the sulfate-reducing bacteria has siroheme like assimilatory sulfite reductase from *Escherichia coli* (Murphy et al., 1973). Four kinds of sulfite reductases have been found in the sulfate-reducing bacteria: desulfoviridin, desulforubidin, desulfofuscidin, and P582. These enzymes differ from each other in the microenvironments where siroheme is present in the molecule (Fauque et al., 1991).

As sulfite reductase is a water-soluble enzyme, previously it has been extracted without the aid of a detergent. The water-extracted enzyme has a molecular mass of 200 kDa (subunits, 55 and 45 kDa); its absorption peaks are at 390, 410, 585, and 628 nm (Kobayashi et al., 1972). Steuber et al. (1994) purified the enzyme with the aid of a detergent from *D. desulfuricans*. The detergent-extracted enzyme shows the absorption peaks at 391, 410, 583, and 630 nm, and its molecular mass is also 200 kDa but consists of three subunits (50, 45, and 11 kDa). The detergent-extracted enzyme catalyzes the reduction of hydrogensulfite with ferrocytochrome

$c_3$ directly to hydrogen sulfide; without using the intermediates trithionate and thiosulfate.

$$\left.\begin{matrix} 3H_2 \\ 6H^+ \end{matrix}\right\} \rightarrow \text{Hase} \rightarrow \text{Cyt } c_3 \rightarrow \text{SiR} \rightarrow \left\{\begin{matrix} HSO_3^- + 7H^+ \\ H_2S + 3H_2O \end{matrix}\right. \qquad (4.1)$$

[Hase: hydrogenase; Cyt $c_3$: cytochrome $c_3$; SiR: sulfite reductase]

Thus, it has been established that cytochrome $c_3$ functions as the electron donor for sulfite reductase.

### (e) Siroheme

Siroheme was first found as the prosthetic group of assimilatory sulfite reductase from *E. coli* (Murphy et al.,1973). The heme is a complex of sirohydrochlorin with iron (Fig. 4.3), and sirohydrochlorin is biosynthesized from uroporphyrinogen III via dihydrosirohydrochlorin (Fig. 4.4).

Dihydrosirohydrochlorin is known to be the intermediate also for the biosynthesis of vitamin $B_{12}$ (Battersby, 1994) and coenzyme $F_{430}$ (Mucha et al., 1985; Thauer and Bonacker, 1994) other than for the biosynthesis of siroheme. Although the intermediate compounds in the biosynthesis of cobyrinic acid, a precursor of vitamin $B_{12}$, from dihydrosirohydrochlorin are fully known, the enzymes which catalyze the respective processes are not yet completely known. Those enzymes have not been discovered that catalyze the processes in the biosynthesis of coenzyme $F_{430}$ from dihydrosirohydrochlorin, though the insertion of nickel to the porphyrin nucleus is known to occur at a relatively earlier stage of the biosynthesis and the rearrangement of side chains occurs after the insertion of nickel (Thauer and Bonacker, 1994). Porphyrin IX is biosynthesized in *D. vulgaris* Miyazaki F also via dihydrosirohydrochlorin (Akutsu et al., 1993).

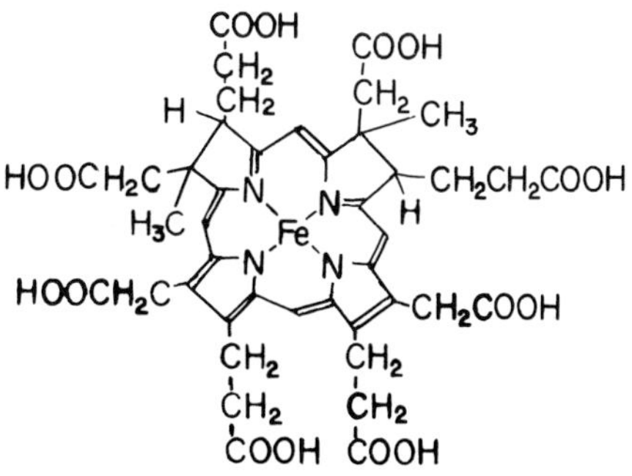


**Fig. 4.3.** Structure of siroheme (Murphy et al., 1973)

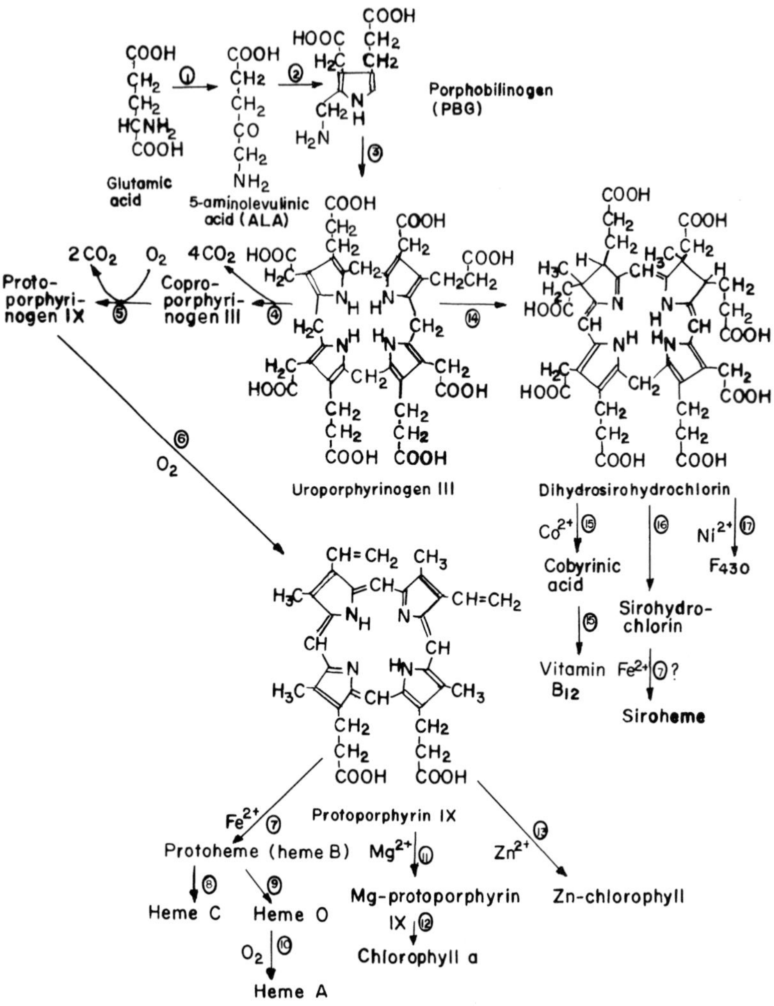


**Fig. 4.4.** Biosynthesis pathways of various metal complexes of porphyrins (including corrin) in plants and many non-photosynthetic bacteria. *Circled numbers*: *1*, Glutamate-tRNA ligase + glutamate-tRNA reductase + glutamate-1-semialdehyde 2,1-aminomutase; *2*, porphobilinogen synthase (ALA dehydrogenase); *3*, hydroxymethylbilane synthase (PBG deaminase) + uroporphyrinogen III synthase (cosynthase); *4*, uroporphyrinogen decarboxylase; *5*, coproporphyrinogen oxidase; *6*, protoporphyrinogen oxidase; *7*, ferrochelatase; *8*, holocytochrome *c* synthase (cytochrome *c* heme-lyase); *9*, protoheme IX farnesyl transferase (Saiki et al., 1992); *10*, oxidation of 8-$CH_3$ of heme O (Svensson et al., 1993; Sakamoto et al., 1999); *11*, (magnesium-) chelatase (Castelfranco et al., 1994); *12*, magnesium-protoporphyrin IX *O*-methyltransferase + magnesium-protoporphyrin IX monomethylester (oxidative) cyclase + protochlorophyllide (4-vinyl) reductase (requiring $NADP^+$ and light) (Castelfranco et al., 1994); *13*, (zinc-) chelatase (zinc-chlorophyll was discovered by Wakao et al., 1996); *14*, uroporphyrinogen III *C*-methyl transferase (requiring *S*-adenosyl-L-methionine) (Warren et al., 1994); *15*, Battersby, 1994; *16*, dehydrogenase ($NAD^+$/$NADP^+$) (Warren et al., 1994); *17*, Mucha et al., 1985; Thauer and Bonacker, 1994). In the higher animals and some bacteria, ALA is synthesized from glycine and succinyl-CoA by the catalysis of ALA synthase

### 4.1.3 Sulfate-Reducing Bacteria and Molecular Oxygen

Although the sulfate-reducing bacteria are strict anaerobes, *Desulfovibrio vulgaris* Miyazaki F has a gene which encodes cytochrome *c* oxidase-like protein (Kitamura et al., 1995). This may mean that the aerobic bacteria might have evolved from the sulfate-reducing bacteria. *Desulfovibrio gigas* accumulates polyglucose under anaerobic conditions and can get energy by fermentation even after it is transferred to the aerobic conditions where sulfate is not available (Santos et al., 1993; Fareleira et al., 1997). Furthermore, some of the sulfate-reducing bacteria grow by oxidizing sulfide under microaerobic conditions (Fuseler et al., 1996; Johnson et al., 1997). Although some bacteria of *Desulfovibrio* genus show a high activity to reduce molecular oxygen with hydrogen, they cannot grow under aerobic conditions. Therefore, the activity is exclusively utilized to delete molecular oxygen (Baumgarten et al., 2001).

Recently, some bacteria have been found which oxidize organic or inorganic compounds with arsenate or selenate (arsenic and selenium respiration) (Stolz and Oremland, 1999). Although many of these bacteria are heterotrophic, *Desulfovibrio auripigmentum* oxidizes hydrogen with arsenate (and with sulfate, sulfite, thiosulfate, and fumarate).

### 4.1.4 Sulfur Respiration

In sulfur respiration, hydrogen and organic compounds are oxidized by elemental sulfur under anaerobic conditions (Pfennig and Biebl, 1976). Sulfur respiration seems to be a variant of sulfate respiration, or rather to be an evolutionary ancestor of the latter. The bacteria that oxidize hydrogen with elemental sulfur are chemolithoautotrophs. Recently, many bacteria which acquire energy by sulfur respiration (sulfur respirers) have been found in the surroundings of hydrothermal vents on the deep sea bottom (Stetter, 1994). Among the chemolithoautotrophic sulfur respirers, cytochromes *b* and *c* have been found. These cytochromes function as the electron carrier between hydrogenase and elemental sulfur reductase (see below). As the lithoautotrophic sulfur respirers seem to be intimately related to the origins of life, the bacteria will be described again later.

A heterotrophic sulfur respirer, *Desulfuromonas acetoxidans*, has been discovered which oxidizes acetic acid and ethanol with elemental sulfur to obtain energy for life processes (Pfennig and Biebl, 1976).

$$\begin{gathered} CH_3COOH + 2H_2O + 4S^{\circ}_{(rhombic)} \rightarrow 2CO_2 + 4H_2S \\ \Delta G^{\circ\prime} = -5.7\,kcal/mol \end{gathered} \quad (4.2)$$

Little is known about elemental sulfur reductase of the sulfur respiration system, though cytochrome $c_3$ of the sulfate-reducing bacteria has been reported to reduce colloidal elemental sulfur in the presence of hydrogen gas and hydrogenase (Fauque et al., 1979). Cytochrome *c*-551.5 (cytochrome $c_7$) (Probst et al., 1977) isolated from *Desulfuromonas acetoxidans* has been found to catalyze the reduction of polysulfide in cooperation with [Fe]hydrogenase from *Desulfovibrio vulgaris* in the presence of hydrogen gas (Pereira et al., 1997). The cytochrome does not catalyze the reduction of colloidal sulfur. Recently, a complex of membrane-bound hydrogenase and sulfur reductase has been purified from hyperthermophilic and acidophilic archaeon, *Acidianus ambivalens* (Laska et al., 2003). The sulfur reductase contains molybdenum. However, this enzyme is not separable without loss of activity from the hydrogenase.

A sulfur reductase has been obtained from a heterotroph, *Wolinella succinogenes* (Schröder et al., 1988). This sulfur reductase has molecular mass of 200 kDa (made up of 85 kDa subunits) and an Fe/S cluster, but no heme. The enzyme catalyzes the oxidation of formate with elemental sulfur in the presence of formate dehydrogenase. As formate dehydrogenase used here has cytochrome *b*, the sulfur-reducing system of the bacterium is said to include cytochrome *b*.

Flavocytochromes *c* from *Chlorobium limicola* f. *thiosulfatophilum* and *Allochromatium* (formerly *Chromatium*) *vinosum* have been reported to catalyze the reduction of elemental sulfur with strong reducing reagents such as benzyl viologen radical (Fukumori and Yamanaka, 1979; Yamanaka and Fukumori, 1980).

### *4.1.5 Autumnal Dying of Rice Plants*

Before around 1940, it was often observed in Japan that the rice plants that had grown well during the summer suddenly died at the beginning of autumn when the plants put forth ears. The phenomenon is called *akiochi* ("autumnal dying of rice plants"). The main cause of the phenomenon is attributable to the action of the sulfate-reducing bacteria. As the rice paddies are covered by a water layer (ca. 3–5 cm in depth), most parts of the soil beneath the water are anaerobic except for several millimeters to one centimeter thickness of the surface of the soil which is aerobic. When ammonium sulfate is supplied to the rice plants as the nitrogenous fertilizer, the rice plants utilize ammonium ion efficiently and sulfate ion is left. This ion is reduced by the sulfate-reducing bacteria in the anaerobic part of the soil of rice paddies to make hydrogen sulfide. While ferrous ion is present in the soil, the ion captures the sulfide ion as ferrous sulfide. However, the amount of hydrogen sulfide formed exceeds the amount of ferrous ion and the roots of the rice plants are damaged by hydrogen sulfide. This is the main cause for the autumnal dying of the rice plants. As the sulfate-reducing bacteria are anaerobes, an introduction of air into the soil of the rice paddies is an effective means of stopping the growth of the bacteria (Shioiri, 1943). Thus intermediate drainage of the rice paddies in midsummer several times rescues the rice plants from the autumnal withering. The drainage in midsummer of the paddies is called *nakabosi* ("intermediate drainage"). The protection of the rice plants from the autumnal dying is one of the important examples of successes that microbiology has brought about for agriculture. Presently, ammonium chloride is used as a nitrogenous fertilizer for rice plants in place of ammonium sulfate to prevent the formation of hydrogen sulfide. So the drainage of the rice paddies is becoming unnecessary. However, chloride ions may bring about some harmful effects to the rice plants. The drainage of the rice paddies is also effective in preventing the evolution of methane from the paddies. When the drainage of the rice paddies is rendered unnecessary by changing the nitrogenous fertilizer from ammonium sulfate to ammonium chloride, the evolution of methane from the rice paddies cannot be prevented by the drainage either.

### *4.1.6 Checking How Old the Origin of Life Is*

The sulfate-reducing bacteria are usually known to be harmful organisms which produce hydrogen sulfide. However, here the author will mention that they offer us a novel clue as to how long ago the origins of life occurred.

Sulfur atoms present in Nature are composed mainly of two isotopes, $^{32}S$ and $^{34}S$. The sulfate-reducing bacteria reduce sulfate containing $^{32}S$ faster than the salt containing $^{34}S$. Therefore, sulfide produced by the action of the bacteria contains more $^{32}S$ than the sulfate from which sulfide is formed. The isotopic ratio, $^{32}S/^{34}S$, in the sulfur compounds without being attacked by the sulfate-reducing bacteria such as those in fresh volcanic rocks and in meteorites, is 22.21. The ratio in the sulfide formed by the action of the bacteria is larger than 22.21. So if we check the isotopic ratios in sulfide contained in the geological strata of various ages, we may be able to find out the upper limit of the age when the bacteria resided on Earth. Even though the bacteria are primitive, they still seem to be evolutionarily much higher organisms than the living things just generated by the origin of life. Therefore, the age of the origin of life will be older than the upper limit of the age when the sulfate-reducing bacteria resided. In Table 4.1, the isotopic ratios of sulfur compounds in various geological strata are listed. It is seen that the ratio becomes larger according to the newness of the stratum, i.e., the ratio of the Cambrian period is 22.37, while that of the Cenozoic era is 23.02, although the ratios do not necessarily increase monotonously with the newness of the stratum. This may be attributable to the sampling conditions of the stratum. In any case, it is recognized that the number of the bacteria present in the Cenozoic era is much larger than that in the Cambrian period.

Now, what is the situation of much older strata than the Cambrian period? The iron ore strata in the Canadian Shield are known to have been formed 2750 million years ago. The isotopic ratio of the sulfide ores in the strata is 22.49, and this ratio suggests that the number of the bacteria which resided in the strata was one tenth of that of the bacteria in the present day. The strata in the Isua area in West Greenland are known to have been formed 3700 million years ago. The isotopic ratio of the sulfide ores in the strata is 22.24. The researchers have concluded that this ratio suggests that the bacteria have not resided in the strata. Therefore, the age of the

**Table 4.1.** Ratios of $^{32}S/^{34}S$ in the strata of various geologic times[a]

| Geologic time division | Years ago (unit: million years) | $^{32}S$ / $^{34}S$ |
|---|---|---|
| Precambrian era | | |
| (West Greenland, Isua area) | 3700 | 22.24 |
| (Canadian Shield) | 2750 | 22.49 |
| Cambrian period | 575–509 | 22.37 |
| Ordovician–Silurian period | 509–416 | 22.35 |
| Devonian period | 416–367 | 22.17[b] |
| Carboniferous period | 367–289 | 22.35 |
| Permian period | 289–247 | 22.17[b] |
| Triassic period | 247–212 | 22.97 |
| Jurassic period | 212–143 | 22.59 |
| Cretaceous period | 143–65 | 22.74 |
| Cenozoic era | 65–present | 23.02 |

[a] Prepared on the basis of Ault and Kulp (1959), Thode (1980)

[b] It may be attributable to, e.g., the sampling of sediment that the figures are smaller than 22.21

origins of life seems to be older than 2750 million years ago but more recent than 3700 million year ago, on the basis of the trace of action of the sulfate-reducing bacteria. Recent studies on the origin of life based on molecular biological findings, traces of microbial actions left on rocks, the ratio of $^{12}C/^{13}C$ in organic compounds, etc. suggest that life originated 3500 million or more years ago.

## 4.2 Sulfur-Oxidizing Bacteria

As described in the preceding sections, two groups are known in the sulfur-oxidizing bacteria; i.e., one oxidizes sulfur compounds such as hydrogen sulfide, elemental sulfur, and thiosulfate utilizing light energy, while the other oxidizes the compounds in the dark. The bacteria of the former group are called photosynthetic sulfur bacteria, and those of the latter group are called (chemolithoautotrophic) sulfur-oxidizing bacteria. The photosynthetic sulfur bacteria include green sulfur bacteria and purple sulfur bacteria. The green sulfur bacteria (e.g. *Chlorobium limicola*) are green in color and strict anaerobes. The bacteria form the particles of elemental sulfur surrounding the cells when they grow on hydrogen sulfide or sodium sulfide. The purple sulfur bacteria (e.g. *Allochromatium vinosum*) are wine red in color and microaerobes. The bacteria form the particles of elemental sulfur inside the cells when they grow on hydrogen sulfide or sodium sulfide. These photosynthetic bacteria oxidize hydrogen sulfide to sulfuric acid when the supply of the sulfide is limited, while the oxidation of the sulfide by them stops at the elemental sulfur stage when sulfide of higher concentrations is continuously supplied.

As the aim of this book is to treat chemolithoautotrophic bacteria, the lithoautotrophic sulfur-oxidizing bacteria will be mainly described in this section. The bacterial group contains many species: *Acidithiobacillus thiooxidans* (formerly *Thiobacillus thiooxidans*), *Thiobacillus neapolitanus*, *Starkeya novella* (formerly *Thiobacillus novellus*), etc. The former two species are strict chemolithoautotrophs and the latter one is a facultative chemolithoautotroph, i.e., this bacterium grows not only on inorganic compounds but also on organic compounds. Their growth pH will be important in relation to the discussion below. There are two groups regarding growth pH (Kuenen, 1989); the bacteria of one group grow at pH 5–8 (e.g., *T. neapolitanus*) (some grow at pH 5–10) and those of the other group grow at pH 1–5 (e.g., *T. thiooxidans*). The bacteria of this group grow even at pH 0.7–6.5 in a special case. The growth pH of these bacteria is important in considering the corrosion mechanisms of concrete by them. Namely, the pH of the surface of fresh concrete is 12–13 and any sulfur-oxidizing bacteria do not seem to grow on such a surface. So it has been said that after the pH of the surface of concrete has been decreased to about 7–8 by the exposure to atmospheric carbon dioxide for 10 or more years (neutralization of concrete), the bacteria of growth pH 5–8 first start growing on the surface of the concrete, and after the environmental pH has decreased to around 5 by the bacterial oxidation of the sulfur compounds, the bacteria of growth pH 1–5 grow actively and the environmental pH decreases further. Many researchers think that the bacteria of growth pH 1–5 cannot grow on the surface of concrete just naturally neutralized. However, the author will show below that this idea is incorrect.

The sulfur-oxidizing bacteria oxidize hydrogen sulfide, elemental sulfur, and thiosulfate to sulfuric acid or sulfate. The sulfur compounds present in Nature most abundantly are hydrogen sulfide, elemental sulfur, and pyrite ($FeS_2$), except for sulfur dioxide. Oxidation by the sulfur-oxidizing bacteria of the former two compounds is most important when the bacterial oxidation of sulfur compounds in natural environments is considered. This is because, though pyrite ($FeS_2$) is also abundantly present in Nature, it is not oxidized easily by the action of the usual sulfur-oxidizing bacteria. The compound is oxidized by the action of the acidophilic iron-oxidizing bacteria (see below).

## 4.2.1 *Bacterial Oxidation Mechanisms of Sulfur Compounds*

The oxidation of sulfur compounds in the sulfur-oxidizing bacteria is performed by the combination of several elemental reaction processes, and each of the processes is performed by a cooperation of several enzymes and cytochromes. The elemental reaction processes included in the bacterial oxidation of sulfur compounds will be described in the following section.

### (a) Oxidation of Sulfide and Elemental Sulfur

Although sulfide (mainly $H_2S$) may be oxidized nonenzymatically to elemental sulfur by giving its electrons to ferricytochrome *c*, sulfide-cytochrome *c* oxidoreductase is found in *Thiobacillus* sp. W5 (Visser et al., 1997c). Furthermore, FMN-containing flavocytochrome *c* is obtained from *Beggiatoa alba* (Schmidt and DiSpirito, 1990). This protein resembles *Chlorobium limicola* and *Allochromatium vinosum* sulfide-cytochrome *c* reductases (flavocytochromes *c*) (Yamanaka, 1994). However, it has not yet been decided whether the enzymes or proteins of *Thiobacillus* sp. W5 and *Beggiatoa alba* participate in the oxidation of sulfide, in vivo. Incidentally, a cyanobacterium, *Microcoleus chthonoplastes*, oxidizes sulfide to thiosulfate (de Wit and Gemerden, 1987), and *Rhodopseudomonas marina* has a sulfide oxidation mechanism similar to that of the cyanobacterium (Imhoff, 1983).

In many bacteria of *Thiobacillus* genus, elemental sulfur is oxidized to sulfite by the catalysis of sulfur dioxygenase. The *Starkeya novella* oxygenase requires GSH (reduced glutathione) (Charles and Suzuki, 1966a) in vitro, while the *Acidianus* (*Sulfolobus*) *brierleyi* enzyme does not require GSH (Emmel et al., 1986). This enzyme has molecular mass of 560 kDa composed of 35 kDa subunits, and has iron as the prosthetic group. Moreover, from *Acidianus* (*Desulfurolobus*) *ambivalens*, sulfur oxygenase reductase is obtained which changes elemental sulfur to sulfide and sulfite simultaneously (Kletzin, 1989).

### (b) Oxidation of Thiosulfate

When thiosulfate and ferricytochrome *c* (native and horse) are added to the cell-free extract of *Starkeya novella*, the ferricytochrome *c* is reduced, and further addition of cyanide does not accelerates the reduction of ferricytochrome *c*. However, when a partially purified cell-free extract of the bacterium is used, the reduction of ferricytochrome *c* with thiosulfate occurs only on addition of cyanide (Fukumori et al., 1989). In this bacterium, thiosulfate-cleaving enzyme (thiosulfate-sulfur transferase, 76 kDa = 2 × 38 kDa), and sulfur-accepting (or sulfur-binding) protein are present, and thiosulfate is cleaved to "S"-sulfur-accepting protein and sulfite [reaction (4.3)] in the organism. When the sulfur-accepting protein is lost by the partial purification of the cell-free extracts, cyanide is required for the cleaving of thiosulfate [reaction (4.4)] (Charles and Suzuki, 1966a,b).

$$S_2O_3^{2-} + \text{sulfur-accepting protein} \xrightarrow{\text{thiosulfate-cleaving enzyme}} \text{"S"-sulfur-accepting protein} + SO_3^{2-} \quad (4.3)$$

$$S_2O_3^{2-} + CN^- \rightarrow SCN^- + SO_3^{2-} \quad (4.4)$$

The enzyme catalyzing the reaction (4.4) is usually called "rhodanese" and is thought to occur widely in various organisms. However, as "rhodanese" requires cyanide of fairly high concentrations (ca. 1 mM) which are not present in normal organisms, the enzyme seems not to occur in Nature. But enzymes occur which have rhodanese activity, e.g., the thiosulfate-cleaving enzyme shows rhodanese activity. Namely, the term "rhodanese" should not be used as the name of an enzyme, but its use should be limited to the enzymatic activity.

It has been claimed that the enzyme having rhodanese activity does not participate in the oxidation of thiosulfate in *Paracoccus* (*Thiobacillus*) *versutus* and the compound is oxidized directly to sulfate in the bacterium by an enzyme complex (Lu and Kelly, 1983a,b) containing manganese (Cammack et al., 1989). Thiosulfate-cytochrome *c* (or ferricyanide) oxidoreductase occurs, which oxidizes thiosulfate to tetrathionate ($S_4O_6^{2-}$) in *T. neapolitanus* (Trudinger, 1961a,b), *T. thioparus* (Lyric and Suzuki, 1970a,b), *T. tepidarius* (Lu and Kelly, 1988a), *T. thiooxidans* (Chan and Suzuki, 1994), *Acidithiobacillus acidophilus* (Meulenberg et al., 1993), and *Thiobacillus* sp. W5 (Visser et al., 1997b). *Acidithiobacillus acidophilus* has tetrathionate hydrolase which hydrolyzes tetrathionate to elemental sulfur (hydrophilic), thiosulfate, and sulfate (de Jong et al., 1997). Tetrathionate is oxidized to sulfate after it has been taken into the cells in *T. tepidarius* (Lu and Kelly, 1988a). Moreover, it has been detected genetically in *Paracoccus denitrificans* GB17 grown autotrophically on thiosulfate that thiosulfate-cytochrome *c* oxidoreductase occurs, which is probably a flavoenzyme (Wodara et al., 1997). In short, various mechanisms seem to occur in the oxidation of thiosulfate by thiobacilli.

### (c) Oxidation of Sulfite

In *Starkeya novella*, sulfite is oxidized to sulfate by the catalysis of sulfite-cytochrome *c* oxidoreductase [reaction (4.5)]. The enzyme catalyzes the reduction with sulfite of not only native ferricytochrome *c*-550 but also horse ferricytochrome *c* and ferricyanide (Charles and Suzuki, 1966b; Yamanaka et al., 1971, 1981b). The enzyme with a molecular mass of 40 kDa has a cytochrome *c*-551 subunit (23 kDa) (Yamanaka et al., 1981b) and molybdenum (Toghrol and Southerland, 1983). Recently, Kappler et al. (2000) has reported that the molecular mass of the enzyme is 46 kDa and has cytochrome *c* subunit of 8.8 kDa.

$$SO_3^{2-} + 2\text{Cyt}\ c^{3+}\text{-550} + H_2O \xrightarrow{\text{Sulfite-Cyt c oxidoreductase}} SO_4^{2-} + 2\text{Cyt}\ c^{2+}\text{-550} + 2H^+ \tag{4.5}$$

[Cyt $c^{3+}$-550 and Cyt $c^{2+}$-550 are ferricytochrome *c*-550 and ferrocytochrome *c*-550, respectively]

Similar enzymes are known to occur in *Paracoccus* (*Thiobacillus*) *versutus* (Lu and Kelly, 1984a,b) and *T. thioparus* (Lyric and Suzuki, 1970b). The *P. versutus* enzyme has a molecular mass of 44 kDa and contains cytochrome *c*-551. The *T. thioparus* enzyme has one atom each of nonheme iron and molybdenum (Kessler and Rajagopalan, 1972). A membrane-bound type sulfite dehydrogenase has been obtained from *Thiobacillus* (*Acidithiobacillus*) *thiooxidans* JCM 7814. The enzyme has the molecular mass of 400 kDa and catalyzes the reduction of horse ferricytochrome *c* with sulfite (Nakamura et al., 1995, 2001). Also from *Paracoccus* (*Thiosphaera*) *pantotrophus* GB17, sulfite dehydrogenase has been obtained. Its molecular mass is 190 kDa (2 × 47 kDa + 2 × 50 kDa) and it has 4 heme C molecules and 1–2 atoms of molybdenum (Quentmeier et al., 2000). Furthermore,

*Acidianus ambivalens* has a sulfite-acceptor oxidoreductase which catalyzes the reduction of quinone with sulfite (Zimmermann et al., 1999). These results will show that sulfite is usually oxidized to sulfate by the catalysis of sulfite-cytochrome *c* oxidoreductase, more generally sulfite-acceptor oxidoreductase in thiobacilli. However, the role of sulfite-cytochrome *c* oxidoreductase is said not to be clear in *P. versutus*, because thiosulfate is directly oxidized to sulfate without the participation of sulfite-cytochrome *c* oxidoreductase in the bacterium (Lu and Kelly, 1984a). Although sulfite-cytochrome *c* oxidoreductase has been shown to participate in the oxidation of sulfite in *S. novella*, as already mentioned, an alternative thiosulfate-oxidizing pathway is also proposed in which the role of sulfite-cytochrome *c* oxidoreductase is not clear (Kappler et al., 2001). Even in the thiosulfate oxidation system in which sulfite-cytochrome *c* oxidoreductase does not participate, such as reported in *P. versutus*, there is a possibility that the oxidoreductase functions in the oxidation of thiosulfate; thiosulfate is oxidized to sulfate by the catalysis of an enzyme complex which contains a thiosulfate-cleaving enzyme, a sulfur-accepting protein, and sulfite-cytochrome *c* reductase that are tightly bound to each other.

The sulfite oxidation pathway other than that mentioned above occurs in some thiobacilli; sulfite reacts with AMP by the catalysis of adenylylsulfate reductase (APS reductase) to form APS and then APS reacts with diphosphate by the catalysis of sulfate adenylyltransferase to produce sulfate and ATP (Lyric and Suzuki, 1970c; Stille and Trüper, 1984), or APS reacts with orthophosphate by the catalysis of sulfate adenylyltransferase (ADP) and ATP is formed from resulting ADP by the catalysis of adenylate kinase (Zimmermann et al., 1999).

$$SO_3^{2-} + AMP + 2Cyt\ c^{3+} \xrightarrow{\text{adenylylsulfate reductase}} APS + 2Cyt\ c^{2+} \quad (4.6)$$

$$APS + PP_i \xrightarrow{\text{Sulfate adenylyltransferase}} ATP + SO_4^{2-} \quad (4.7)$$

$$APS + P_i \xrightarrow{\text{sulfate adenylyltransferase (ADP)}} ADP + SO_4^{2-} \quad (4.8)$$

$$2ADP \xrightarrow{\text{adenylate kinase}} ATP + AMP \quad (4.9)$$

[Cyt $c^{3+}$, Cyt $c^{2+}$, APS, $PP_i$ and $P_i$ are ferricytochrome c, ferrocytochrome *c*, adenylylsulfate, diphosphate, and orthophosphate, respectively]

Adenylylsulfate reductase of *T. thioparus* with the molecular mass of 170 kDa has one molecule of FAD, 8–10 atoms of non-heme iron, and 4–5 atoms of inorganic sulfide in the molecule (Adachi and Suzuki, 1977).

### (d) Cytochrome *c*

*Starkeya novella* cytochrome *c*-550 was first partially purified by Charles and Suzuki (1966b), and thereafter highly purified by Yamanaka et al. (1971, 1991b). The cytochrome functions as the electron acceptor for sulfite-cytochrome *c* oxidoreductase as mentioned above, and as the electron donor for cytochrome *c* oxidase (Yamanaka and Fujii, 1980). It resembles mitochondrial cytochrome *c* in that it reacts with yeast cytochrome *c* peroxidase as fast as mitochondrial cytochrome *c* (Yamanaka and Kimura, 1974), and in the amino acid sequence (Yamanaka et al., 1991b); the similarity of the cytochrome to horse cytochrome *c* in the sequence is 44%.

### (e) Cytochrome *c* Oxidase

Although cytochrome *c* oxidase from *S. novella* is cytochrome $aa_3$-type, its molecular features are very unique (Shoji et al., 1992). In general, cytochrome $aa_3$-type oxidase has two heme A molecules and 2–3 copper atoms in the minimal structural unit, while the *S. novella* oxidase has one heme A molecule and one copper atom in the minimal structural unit constituted of two subunits (32 and 23 kDa). The oxidase occurs as the monomer of the minimal structural unit in the solution containing 0.5% *n*-octyl-D-β-thioglucoside (OTG). The monomer catalyzes the rapid oxidation of ferrocytochrome *c*, its heme A molecule reacts completely with carbon monoxide (CO) under 100% carbon monoxide atmosphere, and its copper shows a big signal at $g = 2.01$ in EPR spectrum; the monomer seems to be a cytochrome $a_3$ having CuA (Shoji et al., 1992).

In the solution containing 0.5% Tween 20, the oxidase occurs as the dimer of the minimal structural unit. One of two heme A molecules in the dimer reacts with CO; the dimer seems to be cytochrome $aa_3$. The dimer reacts with CO even in air like other cytochromes $aa_3$, while the CO complex of the monomer dissociates easily to CO and the oxidized form of the monomeric oxidase in air. The dimeric oxidase shows proton pumping activity, while the monomeric enzyme does not show proton pumping activity (Shoji, 1992). The dimeric oxidase dissociates to the monomeric enzyme on addition of ATP (more than 700 μM) (Shoji et al., 1999). As the proton pumping activity of cytochrome *c* oxidase is related to the biosynthesis of ATP, the dissociation of the dimeric oxidase by ATP seems to be related with the regulation of ATP biosynthesis in the bacterium. In Table 4.2, some properties of monomeric and dimeric species of *S. novella* cytochrome *c* oxidase are compared.

*Starkeya novella* cytochrome *c* oxidase reacts rapidly not only with native cytochrome *c* but also with tuna and yeast cytochromes *c*, while it reacts very slowly with horse and cow cytochromes *c* (Yamanaka and Fukumori, 1977; Yamanaka and Fujii, 1980). Cytochromes *c* which react rapidly with the oxidase have Tyr (46), while cytochromes *c* which react slowly with the oxidase have Phe (46) (Yamanaka and Fukumori, 1978). Thus, human cytochrome *c* reacts with the oxidase more rapidly than horse and cow cytochromes *c*. However, native cytochrome *c* which reacts rapidly with the oxidase has Phe(46).

Cytochrome $aa_3$-type oxidase of thiobacilli has been highly purified from *Acidithiobacillus ferrooxidans* (Kai et al., 1992) besides the oxidase from *S. novella* as described below. Although *Paracoccus versutus* has cytochrome $aa_3$-type oxidase, it has not yet been purified (Kelly, 1989). Cytochrome $cbb_3$-type cytochrome *c* oxidase has been obtained from *Thiobacillus* sp. W5 which resembles *T.*

**Table 4.2.** Effects of ATP on the molecular features of *S. novella* cytochrome *c* oxidase[a]

| Properties | Molecular state | | |
|---|---|---|---|
| | "$a_3$" + "$a_3$" | Tween 20, −ATP → ← OTC or +ATP | $a_3 + a$ |
| Molecular mass (kDa)[b] | 70 | | 160 |
| Heme A (mol/mol enzyme) | 1 | | 2 |
| Cu (atom/mol enzyme) | 1 | | 2 |
| Reactivity with CO of heme A (%) | 100 | | 50 |
| $E_{m,7.0}$ (V) | +0.263 | | +0.180, +0.350 |
| Proton-pumping activity | − | | + |
| [S]–v curve in the oxidation of ferrocytochrome *c* | Hyperbolic | | Sigmoid |

OTC, *n*-octyl-D-β-thioglucoside

[a] Prepared on the basis of Shoji et al. (1992, 1999)

[b] Molecular mass of the oxidase binding to detergents

*neapolitanus* (Visser et al., 1997a). Cytochrome *b* or *o* seems to function as the terminal oxidase also in *Thiobacillus tepidarius* (Kelly et al., 1993). Furthermore, it has been suggested that the terminal oxidase of *Acidithiobacillus ferrooxidans* NB1-3 is a quinol oxidase (Nogami et al., 1997).

### (f) Oxidation Systems of Sulfite and Thiosulfate

Rapid oxygen consumption is observed when sulfite is added to the mixture of sulfite-cytochrome *c* oxidoreductase, cytochrome *c*-550, and cytochrome *c* oxidase highly purified from *S. novella* (Yamanaka et al., 1981b). The oxygen consumption activity per cytochrome *c* oxidase observed with the above mixture is comparable to that observed with the cell-free extracts of the bacterium. Thus, the sulfite oxidase system of the bacterium has been established to be constituted of the three components mentioned above. The finding that the system does not involve cytochrome *b* agrees with the result that the oxidation of sulfite with the membranes from the bacterium is not inhibited by HOQNO (*n*-heptylhydroxyquinoline *N*-oxide) (Oh and Suzuki, 1977a,b; the author and his colleagues have also confirmed the result), though Beffa et al. (1993) claimed that the oxidation of sulfite by the intact cells of *S. novella* DSM 506 was considerably inhibited by HOQNO. As *S. novella* DSM 506 is reported to have two pathways of the thiosulfate oxidation (Kappler et al., 2001) as mentioned already, the pathway other than that mentioned above may contain cytochrome *b* and the oxidation of sulfite by this pathway may be inhibited by HOQNO. Many reports have been published which suggest, on the basis of the inhibition by HOQNO, that cytochrome *b* participates in the sulfite oxidation by *Thiobacillus tepidarius* (Lu and Kelly, 1988a), *Acidithiobacillus thio-*

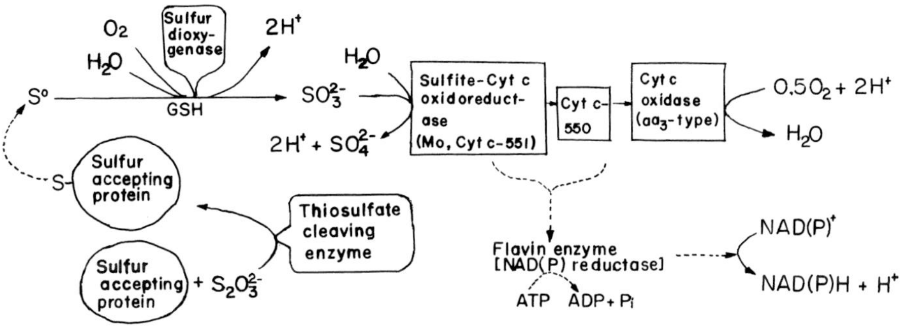


**Fig. 4.5.** The oxidation pathway of sulfur compounds in *S. novella* (Prepared on the basis of Charles and Suzuki, 1966a,b; Aleem, 1966; Yamanaka et al., 1971, 1981b; Yamanaka and Fujii, 1980; Fukumori et al., 1989; Shoji et al., 1992). *Dashed line with arrow*, unverified; *Cyt*, cytochrome; *GSH*, reduced glutathione; *Pi*, phosphate

*oxidans* (Moriarty and Nicholas, 1970; Takakuwa, 1976; Tano et al., 1982; Suzuki et al., 1992), and *Thiobacillus caldus* (Hallberg et al., 1996), though, at least, the sulfite oxidation in *Acidithiobacillus thiooxidans* appears to be catalyzed by sulfite-cytochrome *c* oxidoreductase. Later it was reported that in the case of *T. tepidarius*, the oxidation of sulfite is directly coupled to cytochrome *c*, while the oxidation of tetrathionate formed as the result of the oxidation of sulfite seems to be coupled to cytochrome *b* (Kelly et al., 1993).These results may indicate that the sulfite oxidation pathway varies with species in thiobacilli. However, the sulfite oxidation in these cases has not been studied with a highly purified enzyme system, unlike the case of *S. novella* mentioned above.

On the basis of the results mentioned above, *S. novella* seems to have the oxidation pathway of sulfur compounds as shown in Fig. 4.5, though another pathway may also occur.

In *S. novella*, thiosulfate-cleaving enzyme splits thiosulfate to "S" (S° or $S^{2-}$ bound to sulfur-accepting protein) and sulfite in the presence of a sulfur-accepting protein (Fukumori et al., 1989). Elemental sulfur is oxidized to sulfite by the catalysis of sulfur dioxygenase in the presence of GSH in vitro (Charles and Suzuli, 1966a). When elemental sulfur or sulfide bound to a sulfur-accepting protein is directly oxygenated, probably GSH is not required in the reaction. The sulfite formed is oxidized to sulfate by the catalysis of sulfite-cytochrome *c* oxidoreductase (Charles and Suzuki, 1966b; Yamanaka et al., 1981b).

In *Paracoccus vertusus*, thiosulfate is oxidized directly to sulfate by the catalysis of an enzyme complex containing several cytochromes *c* but not cytochrome *b* (Kelly, 1989). Although sulfite-cytochrome *c* oxidoreductase occurs in the enzyme complex, the enzyme is thought not to participate in the oxidation of thiosulfate, because the rhodanese activity is not observed with the complex. However, as already indicated, it could be that as the enzyme complex contains a thiosulfate-cleaving enzyme strongly bound to both the sulfur-accepting protein and sulfite-cytochrome *c* oxidoreductase, thiosulfate appears to be oxidized directly to sulfate.

Thus, the thiosulfate-cleaving enzyme does not show rhodanese activity in the presence of the sulfur-accepting protein (Fukumori et al., 1989).

Thiosulfate is found to be oxidized to tetrathionate by the catalysis of thiosulfate-cytochrome *c* oxidoreductase in *Acidithiobacillus thiooxidans* JCM 7814 (Nakamura et al., 2001), *Thiobacillus tepidarius* (Lu and Kelly, 1988a) and *Thiobacillus neapolitanus* (Trudinger, 1961b). Tetrathionate is thought to be oxidized in the cytoplasm. But its oxidation mechanism has not been clarified.

*Thiobacillus denitrificans* grows aerobically by the oxidation of thiosulfate with molecular oxygen, while it grows anaerobically by the oxidation of thiosulfate with nitrate. The bacterium is unique in oxidizing thiosulfate with inorganic compound other than molecular oxygen (Adams et al., 1971). In the bacterium, the pathway in the thiosulfate oxidation is similar to that of some other thiobacilli (Aminuddin and Nicholas, 1974b), while the pathway in the nitrate reduction is similar to that of other denitrifiers having cytochrome $cd_1$-type nitrite reductase (Sawhney and Nicholas, 1978). Thiosulfate is first cleaved to sulfite and elemental sulfur, and the sulfite formed is oxidized to sulfate both by the AMP dependent and independent oxidizing systems (Aminuddin and Nicholas, 1974a; Aminuddin, 1980). Both the systems contribute to the biosynthesis of ATP. When the bacterium grows anaerobically on thiosulfate and nitrate, elemental sulfur accumulates transiently and disappears rapidly after thiosulfate has been consumed (Schedel and Trüper, 1980). However, the mechanism by which elemental sulfur is oxidized anaerobically is not known. Recently, *Thioalkalivibrio denitrificans* strain ALJD has been found, which grows anaerobically by the oxidation of thiosulfate and polysulfide ($S_8^{2-}$) with nitrous oxide at pH 10. The bacterium grows also by the oxidation of these electron donors with molecular oxygen (Sorokin et al., 2001). Furthermore, an extreme thermophilic archae bacterium, *Pyrobaculum aerophilum,* has been found, which grows anaerobically by the oxidation of thiosulfate with nitrate (Volkl et al., 1993). A bacterium of *Thioploca* genus is known to grow by the anaerobic oxidation of hydrogen sulfide with nitrate (Jorgensen and Gallardo, 1999).

## 4.2.2 Sulfur-Oxidizing Bacteria Support Animals in the Dark on the Deep-Sea Bottom

Previously, it had been thought that animals can grow only depending on green plants which photosynthesize utilizing solar energy. Therefore, it was a very exciting phenomenon that many animals were found to reside on the dark sea bottom deeper than 2500 m. Around the hydrothermal vents on the deep sea bottom, many mussels, crabs, shrimps, starfishes, and tube worms are present (Grassle, 1985) (Fig. 4.6). Black hot water containing metal sulfides gushes out from the hydrothermal vents, and ores deposit around the vents. The vents are called black smokers or chimneys as they gush out black hot water. Around the hydrothermal vents, sulfur-oxidizing bacteria (or bacterium) of *Beggiatoa* genus grow, creating a mat.

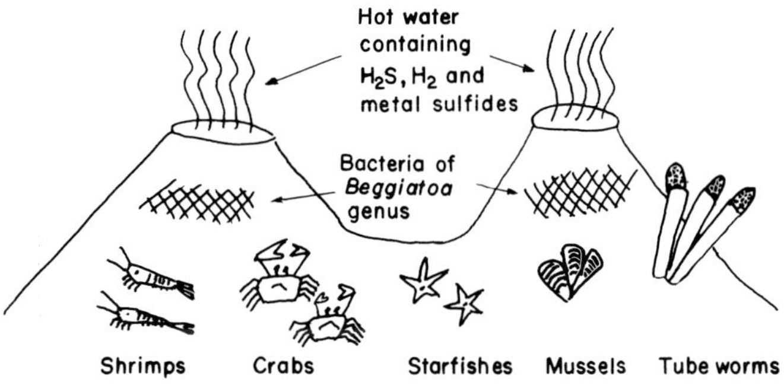


**Fig. 4.6.** A schematic presentation showing that the bacterium (or bacteria) of *Beggiatoa* genus and various animals reside around the hydrothermal vents

The animals such as mussels, crabs etc. feed on the sulfur-oxidizing bacteria (Jannasch et al., 1989).

In the case of tube worms, the situation is a little different from the other animals. The tube worms are approximately 3–5 cm in diameter, and approximately 2 m in length. The worm has a red cap, gill-like organ on the top of the tube. Anus and mouth of the tube worm have been degenerated, and the sulfur-oxidizing bacteria (or bacterium) of *Thiobacillus* genus or rather *Thiovulum* genus on the basis of cellular structures reside in the worm.

Hemoglobin of the worm combines not only with oxygen but also with hydrogen sulfide (Childress et al., 1987). Recently, hemoglobin of another smaller tube worm, the gutless beard worm *Oligobranchia mashikoi*, has been purified and its stereo structure has been determined (Numoto et al., 2005). It has been established on the basis of the structure that the hemoglobin (400 kDa) of the worm has a sulfide-binding site in addition to heme which binds oxygen. The worm utilizes oxygen of oxyhemoglobin by itself, and supplies it also to the bacteria. Furthermore, the worm supplies hydrogen sulfide to the bacteria, the bacteria grow, and the worm feeds them. In short, the worm and the bacteria are in a symbiotic relation.

It was a great surprise that an animal ecology exists which does not depend on photosynthesis but depends on autotrophic chemosynthesis. However, it has been found that molecular oxygen does not come out of the hydrothermal vents. Not only the animals but also the sulfur-oxidizing bacteria need molecular oxygen. Molecular oxygen formed by the photosynthesis of green plants is still utilized by the animals and the bacteria. Thus, the animals on the dark sea bottom cannot reside without the sun. The degree of surprise has thus been reduced by half. However, it is true that the organisms which supply organic compounds to animals are not limited to the photosynthetic organisms but that chemosynthetic autotrophs can also do so. As the sulfur-oxidizing bacteria also need molecular oxygen for their respiration, their growth as well is dependent on sunlight. Recently, many anaerobic chemolithoautotrophs have been found around the hydrothermal vents which acquire energy for their life processes by oxidizing hydrogen gas with elemental

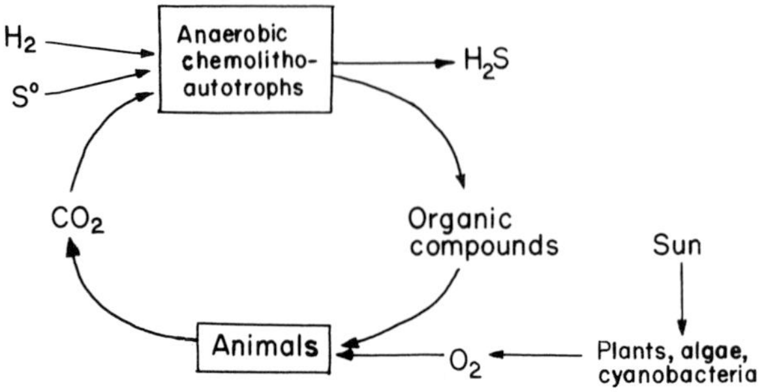


**Fig. 4.7.** A scheme demonstrating that anaerobic chemolitho-autotrophs support animal life on the dark and deep sea bottom

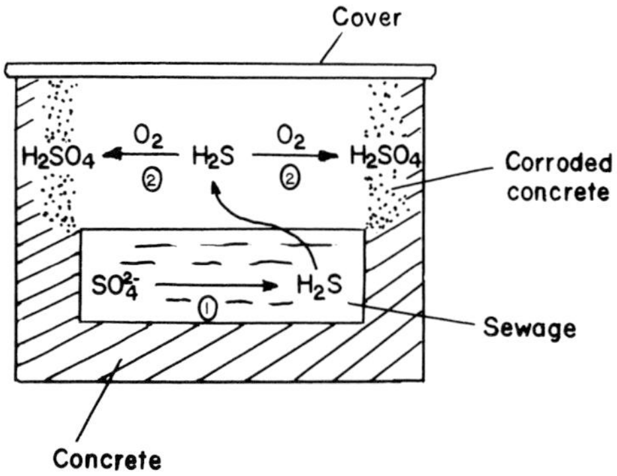


**Fig. 4.8.** A schematic presentation of corrosion of concrete by a cooperation of *1*, the sulfate-reducing bacteria and *2*, the sulfur-oxidizing bacteria in sewerage systems. The surface of concrete dipped in sewage is not corroded

sulfur. When these bacteria support the animal world, the dependency on sunlight of the animals is less than when they grow on the aerobic sulfur-oxidizing bacteria. However, the animals still depend on sunlight because the animals themselves need molecular oxygen (Fig. 4.7).

### *4.2.3 Bacterial Corrosion of Concrete*

Recently, it has been discovered that concrete of sewerage systems is corroded in shorter periods than is expected based on its widely accepted durability. The surface of concrete of the systems becomes friable and the corroded surface can be scratched down by a spatula. The corrosion is caused by a cooperation of sulfate-reducing bacteria and sulfur-oxidizing bacteria (Parker, 1945; Mori et al., 1992). As sewage contains many organic compounds and is anaerobic, the sulfate-reducing bacteria reduce sulfate to hydrogen sulfide therein. The hydrogen sulfide formed goes up into the atmosphere of sewerage systems and is oxidized to sulfuric acid on the surface of the concrete wall. Sulfuric acid formed is concentrated by drying in the micropores of the concrete surface and corrodes the concrete. The corroded part of the concrete is acidic with a pH of 1–2 (Fig. 4.8).

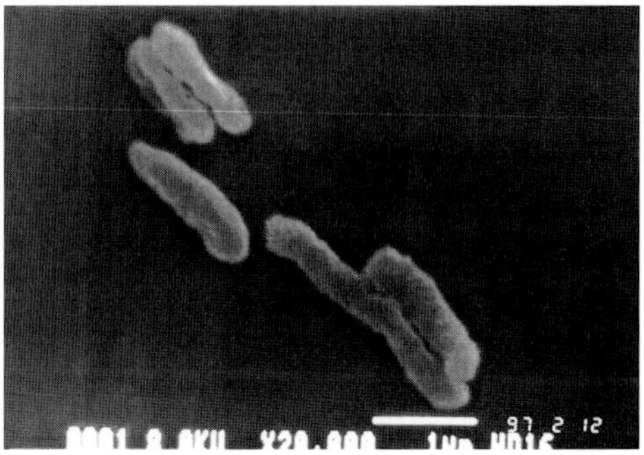

**Fig. 4.9.** Scanning electron microscopic photograph, gold-shadowed, of bacterial cells derived from corroded concrete. *Bar* = 1 μm

When the powder of corroded concrete is mixed with a culture medium for the sulfur-oxidizing bacteria and the mixture aerobically shaken at 28°C, the pH of the medium decreases within days and the pH usually becomes less than 2 after 10 days of shaking. The precipitate collected by centrifugation of the culture medium after 10 days of shaking contains bacteria, as shown in Fig. 4.9.

Many researchers working on concrete have thought that the sulfur-oxidizing bacteria cannot oxidize hydrogen sulfide on the surface of fresh concrete, because the pH of fresh concrete is 12–13. While concrete has been reacting with atmospheric carbon dioxide for a long period, i.e., for 10–15 years, its pH decreases to about neutral (pH 7–8) (neutralization of concrete). It is thought that at this point the bacteria grow on the surface of concrete. As mentioned in the preceding section, there are two groups of sulfur-oxidizing bacteria as regards the growth pH; the bacteria of one group grow at pH 5–8 (or pH 5–10) and those of the other group grow at pH 1–5. Therefore, these researchers think that after the neutralization of concrete, first the bacteria of the former group grow on the surface of the concrete, the pH of the surface decreases to about 5 or less by oxidation of hydrogen sulfide, followed by the bacteria of the latter group that actively oxidize hydrogen sulfide to decrease the pH to 1–2. However, it has recently been found that concrete is corroded much faster than previously expected if the concentration of hydrogen sulfide in the atmosphere of the sewerage systems is as high as ca. 600 ppm (or even if the concentration of the sulfide sometimes reaches 600 ppm) (Yamanaka et al., 2002a). When the test piece of concrete is exposed to about 600 ppm hydrogen sulfide in the atmosphere of a sewerage system, the pH of the surface of the piece becomes 4 two weeks after the exposure has started, and 1 month after the exposure, it becomes 0–1 as checked with a pH test paper. The results suggest that the previous idea about the corrosion of concrete should be modified.

It has become known that sulfur-oxidizing bacteria can grow even when the pH of the surface of the concrete is 12–13, i.e., before the neutralization. How can the bacteria grow on the surface of the concrete at such high pH as 12–13? Where hydrogen sulfide of high concentration is present, the humidity is also high. So a water film containing hydrogen sulfide covers the surface of the concrete. As the pH of saturated hydrogen sulfide solution is about 2, the pH of the water film will be at least less than neutral. Even if the thickness of the film is 0.1 mm ($10^{-4}$ m), it is like a swimming pool for the bacteria of 1–3 μm ($10^{-6}$ m) length and the bacteria will oxidize hydrogen sulfide in the pool regardless of the pH of the surface of the concrete. If the pH of the water film is 5 or less, the two groups of the bacteria, i.e., the bacteria of growth pH 5–8 and 1–5, will grow simultaneously from the start of the time of exposure. As even the bacteria which are said to grow at pH 5–8 can actually grow until pH 2.5, the bacteria of both groups will acidify the water of the film to a pH near 2.5, and the surface pH of the concrete will be decreased to around pH 2.5. Then the bacteria of growth pH 1–5 acidify there further, to pH 1.5 or less. The bacteria which are reported to grow at pH 1–5, can actually grow until pH 0.7 or so. In short, it is expected that all kinds of sulfur-oxidizing bacteria attack concrete from the time when the concrete is exposed to hydrogen sulfide under the conditions where the concentration of hydrogen sulfide is about 600 ppm or more, or its concentration sometimes reaches about 600 ppm.

When the bacteria from the corroded concrete are cultured in a laboratory, the kinds of the bacteria obtained are sometimes different according to the systems from which the corroded concrete is taken. For example, the bacteria derived from the system at an area A grow until final pH 2.5, while those from the system at an area B grow until final pH 1.4. The bacteria from the two areas differ from each other on the basis of G + C content of their DNA. Namely, for example, the G + C content of DNA from the bacteria of area A is 55.8 mol%, while the content of DNA from the bacteria of area B is 51.7–53.6 mol%. On the basis of the G + C content, the kind of the bacteria from the two areas are distinctly different from each other, and it should be said that one species of sulfur-oxidizing bacteria is residing in the corroded concrete at the final stage of corrosion in each area. In the above example, the bacterium from area A seems to be *Thiobacillus neapolitanus* (although its growth pH is reported to be 4.5–8.5, it actually grows until pH 2.5), while the bacterium from area B seems to be *Acidithiobacillus thiooxidans* (although its growth pH is reported to be 1–5, it actually grows until pH 0.7) (cf. Kuenen, 1989). These results show that the bacteria of the growth pH 5–8 also can be predominantly present in the corroded concrete at the final stage of corrosion, depending on the environmental conditions in which the bacterium survives until the final stage of corrosion.

In the corroded concrete, there also reside acidophilic iron-oxidizing bacteria (see below) besides the usual sulfur-oxidizing bacteria. The acidophilic iron-oxidizing bacteria show optimal growth pH at 2.0, and oxidize not only ferrous iron but also sulfur compounds. Therefore, they can participate in the corrosion of concrete. We have to consider the action of both the usual sulfur-oxidizing bacteria and the acidophilic iron-oxidizing bacteria when we exploit the compounds which inhibit the bacterial corrosion of concrete.

Are there any methods to inhibit the growth of the bacteria participating in the corrosion of concrete? A few compounds have been exploited which inhibit the growth of the sulfur-oxidizing and acidophilic iron-oxidizing bacteria. A reagent which contains nickel sulfate and nickel oxide supplemented with a compound of tungsten (probably its oxide) has been exploited [anti-bacterial reagent for concrete, RCF-95 (NMB Co. Ltd, Japan, 1997)]. The reagent is effective on the sulfur-oxidizing bacteria but its effect on the acidophilic iron-oxidizing bacteria is questionable. Another reagent is a silver- and copper-carrying zeolite (a ceramic). When the material is added to the culture medium at the concentration of 500 ppm, the growth of the sulfur-oxidizing bacteria is inhibited (Kurihara, 1999). The material inhibits the growth of sulfate-reducing bacteria and *Bacillus subtilis* at the concentrations of 100 ppm and 5000 ppm, respectively. It has not been reported whether or not the material inhibits the growth of the acidophilic iron-oxidizing bacteria. As the acidophilic iron-oxidizing bacteria are considerably tolerant to heavy metal ions (see p. 90), the heavy metal ions which inhibit the growth of the usual sulfur-oxidizing bacteria might not inhibit that of the iron-oxidizing bacteria at the same concentrations as are effective in inhibiting the growth of the usual sulfur-oxidizing bacteria. As the growth of the sulfur-oxidizing bacteria is inhibited when the microparticles of the silver- and copper-zeolite are added to the culture medium, silver and copper ions seem to intrude into the medium. This means that when the zeolite is used by mixing it with concrete, silver and copper are dissolved from the carrier zeolite as their ions, and spread in the environment. When we exploit the materials which inhibit the growth of the bacteria, we have to bear in mind whether or not the materials are dissolved into the environment and whether or not the dissolved materials are harmful. Recently, it has been reported that tungstate effectively inhibits the growth of acidophilic iron-oxidizing bacteria (Sugio et al., 2001). Although the compound may be expected to be used as an antibacterial agent to protect concrete from bacterial corrosion, it can also pollute the environment, as it is the soluble compound of the heavy metal.

The author and his colleagues have found that calcium formate completely inhibits the growth of the sulfur-oxidizing bacteria and acidophilic iron-oxidizing bacteria at 50 mM (6500 ppm) or more in laboratory experiments (Yamanaka et al., 2002a). Calcium formate is hardly harmful and as it will finally become calcium carbonate, it is expected that the compound will barely pollute the environment.

Calcium formate at 50 mM inhibits completely the growth of the sulfur-oxidizing and the acidophilic iron-oxidizing bacteria when these bacteria are cultured in the laboratory. However, the effect of the formate is not so complete when its effect is checked using mortar (concrete without small stones) test pieces in the sewerage systems. When a mortar test piece of $4 \times 4 \times 16$ cm (ca. 570 g in weight, made of cement, sand, and water) and the test piece containing calcium formate (aqueous solution of calcium formate is used in place of water) are kept by hanging them in the atmosphere of the sewerage systems containing 600 ppm hydrogen sulfide (0.6 ml hydrogen sulfide in 1 liter air), the difference in the degree of their corrosion is distinct between the test pieces without and with calcium formate, though calcium formate does not protect the test pieces completely from corrosion. In the case of exposure to 600 ppm hydrogen sulfide, the surfaces of the test pieces become friable and are easily scratched away by a spatula. After scratching away the corroded surface, the weight of the remaining portion is 45 g and 149 g for the

**Table 4.3.** Some properties of mortar test pieces exposed to hydrogen sulfide in a sewerage system[a]

| Calcium formate added (mM)[b] | Expected concentration of calcium formate as a whole (mM) | pH of washed water | Weight (g) | |
|---|---|---|---|---|
| | | | Before exposure | Remainder after washing |
| 0 | 0 | 2.42 | 572 | 47 |
| 1000 | 111 | 3.73 | 566 | 94 |
| 2000 | 222 | 4.41 | 570 | 149 |

[a]From Yamanaka et al. (2002a)

[b]Concentration of formate in the aqueous portion when the test pieces were prepared

test pieces without the formate and with 2.9% (222 mM) formate, respectively (Table 4.3). Namely, an addition of 2.9% calcium formate protected the test piece from corrosion threefold or more than without the formate, though the formate at the concentration of 2.9% does not completely protect the test piece from corrosion. In any case, as the antibacterial effect of calcium formate is observed also in the exposure of the test piece, the formate is expected to be useful in protecting the concrete of sewerage systems from corrosion. An agent containing calcium formate for inhibiting bacterial growth in concrete is sold as "Nonbacter" by Fujita Corporation, Tokyo, Japan.

# Chapter 5
# Oxidation and Reduction of Iron by Bacteria

## 5.1 Bacteria That Oxidize or Reduce Iron

In Nature, there reside bacteria that acquire the energy for life processes by oxidizing ferrous ion to ferric ion. *Acidithiobacillus ferrooxidans* and *Leptospirillum ferrooxidans* oxidize ferrous ion to ferric ion at pH 2.0 and are the most well known among the bacteria that oxidize ferrous ion. Ferrous ion is easily oxidized spontaneously by molecular oxygen at neutral pH, while at pH 2.0 it is not oxidized spontaneously but is easily oxidized even at pH 2.0 by the action of the acidophilic iron-oxidizing bacteria. These bacteria are utilized in various industrial processes because of their ability to oxidize ferrous to ferric ions at pH 2.0. In particular, *A. ferrooxidans* has been well studied industrially as well as scientifically.

Among the iron-oxidizing bacteria, some acquire the energy for life processes by oxidizing ferrous ion at neutral pH. The best known among these is *Gallionella ferruginea*, which resides in a cluster of ferric hydroxide in the field when it lives on a small scale. As mentioned above, at neutral pH ferrous ion is easily oxidized spontaneously by air (molecular oxygen), so it seems hard for the bacterium to utilize ferrous ion to acquire the energy for life processes at neutral pH. It is probable that on the inside of the ferric hydroxide crust where it lives, partial pressure of oxygen becomes so low that the spontaneous oxidation of ferrous ion occurs very slowly. Therefore, the bacterium utilizes a sufficient amount of ferrous ion to acquire the energy. Thus, the bacterium is cultivated in the laboratory under a reduced partial pressure of molecular oxygen; under one hundredth of the partial pressure in air (Hanert, 1989). On the basis of the calculation, the free energy liberated per ferrous ion at pH 7.0 (ca. 20.8 kcal/$Fe^{2+}$) is about 2.6 times as much as that liberated per ion at pH 2.0 (ca. 8.1 kcal/$Fe^{2+}$). In any case, it is known that the bacterium resides, e.g., on the bottom of Crater Lake, Oregon, USA, making a mat (Dymond et al., 1989). Moreover, some iron-oxidizing bacteria are known which do not need molecular oxygen for the oxidation of ferrous ion. The brief properties of several iron-oxidizing bacteria are listed in Table 5.1.


doi: 10.1007/978-4-431-78541-5_5, 

**Table 5.1.** Various bacteria oxidizing ferrous ion[a]

| Bacteria | Properties | | | | | |
|---|---|---|---|---|---|---|
| | GC content of DNA (mol%) | Reaction in oxidation of iron | Temperature of growth (°C) | pH of growth | Remarks | |
| *Acidithiobacillus ferrooxidans* | 56–59 | $2Fe^{2+} + 2H^+ + 0.5O_2 \rightarrow 2Fe^{3+} + H_2O$<br>$FeS_2 + 3.5O_2 + H_2O \rightarrow Fe^{2+} + 2SO_4^{2-} + 2H^+$ | 10–37 | 1.3–4.5 | | (1) |
| *Leptospirillum ferrooxidans* | 54–56 | $2Fe^{2+} + 2H^+ + 0.5O_2 \rightarrow 2Fe^{3+} + H_2O$ | 30 | 1.7 | | (2) |
| *Gallionella ferruginea* | 54.6 | $2Fe^{2+} + 4H_2O + O_2 \rightarrow 2Fe(OH)_3 + 2H^+$<br>$2FeS + 3H_2O + 1.5O_2 \rightarrow Fe(OH)_3 + 2S°(?)$ | 20–25 | 6.3–6.6 | Microaerobic | (3) |
| Photosynthetic iron-oxidizing bacterium (*Rhodobacter* sp. *or Chromatium* sp.) | ND[b] | $4FeCO_3 + 10H_2O \rightarrow 4Fe(OH)_3 + [CH_2O]^c + 3HCO_3^- + 3H^+$<br>$4FeS + 9HCO_3^- + 10H_2O + H^+ \rightarrow 4Fe(OH)_3 + 4SO_4^{2-} + 9[CH_2O]$<br>$4Fe^{2+} + CO_2 + 11H_2O \rightarrow 4Fe(OH)_3 + [CH_2O] + 8H^+$ | 18–20 | 6.0–7.5 (strain L7)[d]<br>5.5–7.0 (strain SW2)[e] | | (4) |
| *Chlorobium ferrooxidans* | ND | $17FeCO_3 + 29H_2O \rightarrow 17Fe(OH)_3 + [C_4H_7O_3]^c + 13CO_2$ | 18–30 | 6.3–7.1 | Anaerobic[f] | (5) |
| *Ferroglobus placidus*[g] | 43 | $2FeCO_3 + NO_3^- + 6H_2O \rightarrow 2Fe(OH)_3 + NO_2^- + 2HCO_3^- + 2H^+ + H_2O$ | 65–95 | 7 | Archaeon | (6) |

(1) Kuenen (1989); (2) Eccleston et al. (1985); (3) Kucera and Wolfe (1957), Hanert (1989), Hallbeck and Pedersen (1991); (4) Ehrenreich and Widdel (1994); (5) Heising et al. (1999); (6) Hafenbradl et al. (1996)

[a] Besides those listed in the table, a moderately thermophilic and acidophilic iron-oxidizing bacteriun is known which is deficient in the enzymes participating in carbon dioxide fixation (Sugio et al., 1995; probably Brierley and Lockwood, 1977)

[b] Not determined

[c] Organic compounds biosynthesized

[d] Sulfide does not act as electron donor. Grows under nitrogen atmosphere. Light is necessary

[e] When ferrous ion is an electron donor, atmosphere of $N_2$–$CO_2$ (90 : 10, v/v) is used. Light is necessary

[f] For growth by photosynthetically oxidizing ferrous ion, co-cultivation with bacterium is necessary which seems to be *Geospirillum* sp. Light is necessary

[g] This bacterium also oxidizes hydrogen gas and sulfide with nitrate, and hydrogen gas with thiosulfate

### 5.1.1 Mechanisms in Bacterial Oxidation of Iron

Bacterial oxidation of ferrous ion in *Acidithiobacillus ferrooxidans* occurs by the catalysis of a system consisted of several enzymes and proteins. In the oxidation of ferrous ion by *A. ferrooxidans* Fe1 (JCM 7811), electrons are first pulled out of the ion by the catalysis of Fe(II)-cytochrome *c* oxidoreductase. Then, electrons are transferred to ferricytochrome *c*-552 (native cytochrome *c* of the bacterium), ferrocytochrome *c*-552 formed is oxidized with oxygen by the catalysis of cytochrome *c* oxidase (Yamanaka and Fukumori, 1995). However, the mechanism of the oxidation of ferrous ion appears to be a little different between the strains of *A. ferrooxidans*. Thus, in certain strains of the bacterium the oxidation of ferrous ion is catalyzed by Fe(II)-rusticyanin oxidoreductase (Blake and Shute, 1994). However, as will be pointed out below, the enzyme should be carefully checked. Moreover, in a moderately thermophilic iron-oxidizing bacterium, the oxidation of ferrous ion is reported to be catalyzed by an iron oxidase containing heme A (Takai et al., 1999, 2001).

#### (a) Fe(II)-Cytochrome *c* Oxidoreductase

The enzyme has the molecular mass of 63 kDa and is composed of eight 6.2 kDa subunits, each of which has a $[Fe_4S_4]$ cluster (Fukumori et al., 1988b; Yano, 1992). When the enzyme is kept for 2 weeks in a refrigerator, it dissociates to inactive subunits with molecular mass of 26 kDa, and when it is treated at 100°C for 5 min in the presence of SDS, it dissociates to subunits with molecular mass of 6.2 kDa. As the enzyme is obtained in the reduced state when it is purified and its $E_{m,2.0}$ is +0.633 V, it resembles high-potential iron–sulfur protein [HiPIP] from the purple phototrophic bacteria (Tedro et al., 1979, 1985). Thus, the amino acid sequence of the subunit deduced from the base sequence of DNA which encodes the enzyme is very similar to those of HiPIP (Kusano et al., 1992). Therefore, some researchers call the enzyme HiPIP. The finding of an iron–sulfur protein-type enzyme such as Fe(II)-cytochrome *c* oxidoreductase agrees with the report that a protein having an iron–sulfur cluster occurs in *A. ferrooxidans* as the EPR detectable electron transport component on the reducing site of rusticyanin (Fry et al., 1986). On the contrary, Appia-Ayme et al. (1999) have reported that HiPIP does not participate in the oxidation of ferrous ion in *A. ferrooxidans* ATCC 33020 based on DNA studies. Bruscella et al. (2005) claimed that the location of the HiPIP in the electron transfer chain in the bacterium was still ambiguous, so that its physiological role was not clarified.

Fe(II)-cytochrome *c* oxidoreductase catalyzes the reduction of ferricytochrome *c*-552 with ferrous ion at pH 3.5 in vitro, but it does not catalyze the reduction of rusticyanin (a copper protein; see below) with ferrous ion. However, it catalyzes the reduction of rusticyanin with ferrous ion in the presence of a catalytic amount of cytochrome *c*-552 (Yamanaka et al., 1991a). Although the enzyme is extremely labile in the presence of cytochrome *c*-552 in vitro, it is protected by rusticyanin from the inactivation by cytochrome *c*-552. One mole of the enzyme catalyzes the reduction of 98.1 mol of rusticyanin at pH 3.5 in the presence of a catalytic amount of cytochrome *c*-552. Thus, the following electron transfer occurs (Sato et al., 1989).

$Fe^{2+}$→Fe(II)-cytochrome *c* oxidoreductase→cytochrome *c*-552→rusticyanin

The Fe(II)-cytochrome *c* oxidoreductase catalyzes the reduction of cytochrome *c*-552 with ferrous ion but does not catalyze the direct reduction of rusticyanin. This agrees with the finding of Hazra et al. (1992) that rusticyanin is reduced via cytochrome *c*. Blake and Shute (1994) have claimed that the oxidation of ferrous ion occurs by the catalysis of Fe(II)-rusticyanin oxidoreductase. However, as their enzyme preparation probably contained cytochrome *c* as a contaminant, its catalysis could be the reduction of rusticyanin by Fe(II)-cytochrome *c* oxidoreductase mediated by cytochrome *c*. It has been reported that rusticyanin is not present in *Leptospirillum ferrooxidans* (Blake et al., 1993). This may mean that rusticyanin is not necessarily required for the oxidation of ferrous ion by the iron-oxidizing bacteria.

### (b) Cytochromes *c*

Many kinds of cytochromes *c* occur in *Acidithiobacillus ferrooxidans.* One of them is water-soluble cytochrome *c*-552(s) (14 kDa) (hereafter "s" means soluble, and "m" means membrane bound) highly purified by Sato et al. (1989). This cytochrome seems to be the same protein as that partially purified by Vernon et al. (1960) and Ingledew (1982). Besides the cytochrome, several *c*-type cytochromes are obtained: cytochrome *c*-552(m) (22.3 kDa), cytochrome *c*-550(m) (51 kDa) (Tamegai et al., 1994); Valkova-Valchanova and Chan, 1994) and cytochrome *c*-552(m) (30 kDa) (Elbehti and Lemesle-Meunier, 1996), cytochrome $c_4$ (21.2 kDa) (Cavazza et al., 1996), and brown soluble cytochrome *c*-553 (12789 Da) (Cavazza and Bruschi, 1995). Furthermore, occurrence of several cytochromes *c* besides those mentioned above has been indicated (Yarzabal et al., 2002a). It has also been reported that cytochrome *c* (46 kDa) occurs in the outer membrane of the bacterium and participates in removing electrons from insoluble iron compounds such as pyrite (Yarzabal et al., 2002b).

A cytochrome having the α peak at 579 nm is obtained from *Leptospirillum ferrooxidans* DSM 2706 (Hart et al., 1991). The cytochrome has 1 atom of zinc in addition to heme iron. Its molecular mass is 17.9 kDa and its $E_{m,3.5}$ is +0.68 V (Blake et al., 1993). From *L. ferrooxidans* P3A, a cytochrome with the molecular mass of 12 kDa is obtained, while the 17.9 kDa cytochrome is not obtained. In any case, nothing is known about the function of the *L. ferrooxidans* cytochromes. Moreover, *Metallosphaera sedula* and *Acidianus brierleyi* have a membrane-bound yellow cytochrome which is reduced by ferrous ion (Blake and McGinness, 1993; Blake et al., 1993). However, the description here about cytochromes *c* will be limited to the proteins that have been purified from *A. ferrooxidans* and well characterized.

Cytochrome *c*-552(s) (14 kDa) of *Acidithiobacillus ferrooxidans* acts not only as the electron acceptor for Fe(II)-cytochrome *c* oxidoreductase but also as the electron donor for cytochrome *c* oxidase. Cytochromes *c*-552(m) (22.3 kDa) and *c*-550(m) (51 kDa) also act as the electron donor for cytochrome *c* oxidase although it has not yet been clarified whether they act as the electron acceptors for Fe(II)-cytochrome *c* oxidoreductase. The reactivity with cytochrome *c* oxidase of cytochrome *c*-550(m) is larger than that of cytochrome *c*-552(s), while the reactivity of cytochrome *c*-552(m) is much less than that of cytochrome *c*-552(s) (Kai et al., 1992; Yamanaka and Fukumori, 1995). However, a big difference is observed in the effect of sulfate on the reactions with the oxidase between these cytochromes *c*; the reaction with the oxidase of cytochrome *c*-552(s) is much stimulated by sulfate ion, while those of cytochromes *c*-552(m) and *c*-550(m) are inhibited by the ion. Considering that *Acidithiobacillus ferrooxidans* requires sulfate for its growth (Lazaroff, 1963), the stimulation of the reaction with the oxidase of cytochrome *c* by sulfate ion seems to suggest that cytochrome *c*-552(s) is a real electron donor for cytochrome *c* oxidase, although several kinds of cytochromes *c* occur in the bacterium (cf. Fig. 5.2).

The amino acid sequences of both cytochromes *c*-552(s) (14 kDa) and *c*-552(m) (22.3 kDa) (Yano, 1992; Yamanaka and Fukumori, 1995) resemble those of cytochromes $c_2$ and $c_4$ (Meyer and Kamen, 1982). Cavazza et al. (1996) purified cytochrome $c_4$ (21.2 kDa), which has two heme C molecules in the molecule. Cytochrome *c*-552(m) (22.3 kDa) seems to be identical with cytochrome $c_4$ (21.2 kDa) on the basis of the amino acid sequence and molecular mass. The author and his colleagues were unable to find two molecules of heme C in the cytochrome *c*-552(m) (22.3 kDa) molecule. Giudici-Orticoni et al. (2000) reported that cytochrome $c_4$ or two-heme cytochrome *c*-552(s) is intimately related to the oxidation of ferrous ion by rusticyanin. The cytochrome seems to be similar to the cytochrome *c* included in ferrous rusticyanin oxidoreductase (Blake and Shute, 1994). Furthermore, Appia-Ayme et al. (1998) claimed that as DNA that encodes cytochrome *c*-552(s) (14 kDa) was not found, this cytochrome could be a proteolytic product of cytochrome $c_4$. However, they have not produced a distinct result showing that cytochrome $c_4$ functions as the electron donor for cytochrome *c* oxidase like cytochrome *c*-552(s) (14 kDa). The reactivity with cytochrome *c* oxidase of cytochrome *c*-552(m) (22.3 kDa) or cytochrome $c_4$ is very low in comparison to cytochrome *c*-552(s) (14 kDa) and its reaction with the oxidase is depressed by sulfate while that of cytochrome *c*-552(s) (14 kDa) is much accelerated by the salt, as already mentioned (Kai et al., 1992; Yamanaka and Fukumori, 1995). These results seem to argue against cytochrome *c*-552(s) (14 kDa) being a proteolytic product of cytochrome *c*-552(m) (22.3 kDa)or cytochrome $c_4$.

### (c) Rusticyanin

Rusticyanin is a copper-containing blue protein. It was first purified by Cox and Boxer (1978). The author and his colleagues purified it to a homogeneous state and determined its amino acid sequence (Yano et al., 1991). In the same year, Ronk et al. (1991) determined the sequence of the protein, and 2 years later Nunzi et al. (1993) also determined its sequence. The sequences of the protein determined with different preparations in the three laboratories are almost identical, although the bacterial strains used seem to be different among the laboratories. The protein molecule is composed of 155 amino acid residues. When it is biosynthesized it has a signal peptide consisting of 32 amino acid residues which are cleaved away when the protein is taken into the periplasm of the bacterium (Bengrine et al., 1998). The reason rusticyanin is stable at pH 2.0 seems explainable by the finding that copper is located in a very hydrophobic environment in the protein molecule (Botuyan et al., 1996). Rusticyanin has been crystallized and its structural group is $P2_12_12_1$ (Djebli et al., 1992).

As described already, although rusticyanin was previously thought to be an iron-oxidizing enzyme (Cobley and Haddock, 1975), Blake and Shute (1987) have demonstrated that the reduction rate of rusticyanin with ferrous ion is too slow to explain the oxidation rate of the ion by *Acidithiobacillus ferrooxidans* cells. Thus, rusticyanin is not present in the acidophilic iron-oxidizing bacterium, *Leptospirillum ferrooxidans* as already mentioned (Blake et al., 1993). Moreover, the protein is present not only in the cells of *A. ferrooxidans* grown on ferrous ion but also in the cells grown on elemental sulfur or thiosulfate at the same concentration as in those grown on ferrous ion (Bengrine et al., 1998). Therefore, rusticyanin seems not inevitable to the oxidation system of ferrous ion in *A. ferrooxidans*. The physiological role of rusticyanin in the bacterial cells will be mentioned below in the section on the iron-oxidizing system.

### (d) Cytochrome *c* Oxidase

Cytochrome *c* oxidase of *Acidithiobacillus ferrooxidans* was previously called cytochrome $a_1$, as it shows the α peak at 595 nm (Ingledew, 1982). However, as the oxidase purified from the bacterium has two heme A molecules and two copper atoms in the minimal functional unit and one of the two molecules of heme A combines with carbon monoxide, it is a cytochrome $aa_3$-type cytochrome *c* oxidase although it has the α peak at 595 nm (Kai et al., 1992). It differs from the usual cytochrome $aa_3$ in having only one molecule of heme A and one atom of copper in the minimal structural unit, which comprises one molecule each of three kinds of subunits (54 kDa, 21 kDa, 15 kDa) like *Starkeya novella* cytochrome *c* oxidase (Shoji et al., 1992). [The DNA study suggests the presence of four subunits with the molecular masses of 69, 28, 18 and 6.4 kDa (Appia-Ayme et al., 1999)]. The minimal functional unit of the *A. ferrooxidans* oxidase is a dimer of the minimal structural unit, and the dimer shows general properties of cytochrome $aa_3$ except that the α peak is present at a wavelength shorter than 600 nm of the absorption spectrum. The oxidase resembles *Nitrosomonas europaea* cytochrome *c* oxidase (Yamazaki et al., 1985) (see pp. 25–26) in the position of the α peak of the absorption spectrum.

The *A. ferrooxidans* oxidase catalyzes the oxidation of ferrocytochrome *c*-552(s) (14 kDa), ferrocytochrome *c*-552(m) (22.3 kDa) or ferrocytochrome $c_4$, and ferrocytochrome *c*-550(m) (51 kDa). Although the oxidase also catalyzes the oxidation of the reduced form of rusticyanin, its $K_m$ for rusticyanin is fairly large (600 μM) (Kai et al., 1992; Yamanaka and Fukumori, 1995) while $K_m$s of the oxidase for cytochromes *c*-552(s), *c*-552(m) and *c*-550(m) are 17, 2.2, and 4.2 μM, respectively. However, as the concentration of rusticyanin in the bacterial cells is considerably high, the copper protein can also be the electron donor for the oxidase in vivo. Although the oxidase shows a catalytic capability to oxidize molybdenum blue (Sugio et al., 1992a), the catalysis is not limited to the *A. ferrooxidans* oxidase because some oxidases belonging to cytochrome $aa_3$ also show such activity (Kai et al., 1992).

One of the characteristics of the oxidase is acidic optimal pH in the catalytic activity; the oxidase shows maximal activity at pH 3.5–4.0 in vitro (Kai et al., 1989, 1992). The finding that the catalytic oxidation rate by the oxidase of ferrocytochrome *c*-552(s) (14 kDa) is maximal at pH 3.5 in vitro suggest that the oxidase in vivo shows its maximal activity round pH 2.0 where the bacterium grows. Namely, the oxidase could reduce molecular oxygen in the periplasmic space but not in the cytoplasm (see Fig. 5.2, p. 87). Furthermore, the oxidation of ferrocytochrome *c*-552(s) (14 kDa) catalyzed by the oxidase is greatly stimulated by sulfate ion, and in particular, the oxidation catalyzed by the oxidase of reduced rusticyanin scarcely occurs in the absence of the ion (Kai et al., 1992). The stimulation by sulfate of the catalytic activity of the oxidase may be related to the result that the bacterium requires sulfate for its growth (Lazaroff, 1963). If so, the physiological electron donor for the oxidase is thought to be cytochrome *c*-552(s) (14 kDa) among the three kinds of cytochromes *c* mentioned above, and rusticyanin could be also the physiological electron donor for the oxidase. Cytochrome *c* oxidase is also highly purified from *A. ferrooxidans* OK1-50 and AP19-3 (Iwahori et al., 1998). Its properties are similar to those of the oxidase from *A. ferrooxidans* Fe1 (JCM 7811). Furthermore, cytochrome *bd*-type quinol oxidase has been purified from *A. ferrooxidans* NASF-1 (Kamimura et al., 2001). A new iron oxidase of cytochrome *a*-type is obtained from a moderately thermophilic iron-oxidizing bacterium. The oxidase shows the α peak at 602 nm and catalyzes the oxidation of ferrous ion at pH 3.0 and at 45°C (Takai et al., 2001). The oxidase will be mentioned again below in relation to the oxidation of ferrous ion by the iron-oxidizing bacteria.

### (e) Electron Transfer System Coupled to Oxidation of Ferrous Ion

In the oxidation of ferrous ion by *A. ferrooxidans* Fe1 (JCM 7811), first, Fe(II)-cytochrome *c* oxidoreductase catalyzes the reduction of ferricytochrome *c*-552(s) with ferrous ion, while it catalyzes the reduction of rusticyanin with ferrous ion only in the presence of a catalytic amount of cytochrome *c*-552(s) (14 kDa) (Fukumori et al., 1988b; Yamanaka and Fukumori, 1995). Although Fe(II)-cytochrome *c*

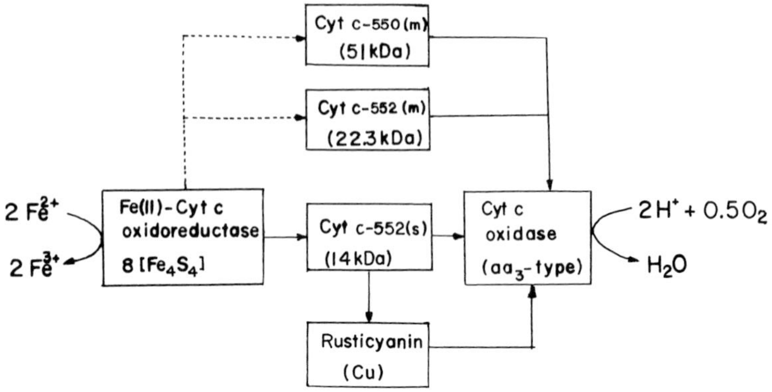


**Fig. 5.1.** Electron transfer pathway in the oxidation of ferrous ion by *Acidithiobacillus ferrooxidans* Fe1 (JCM 7811) (prepared on the basis of: Fukumori et al., 1988b; Sato et al., 1989; Kai et al., 1992; Tamegai et al., 1994). *Dashed line with arrow*, unverified; *Cyt*, cytochrome

oxidoreductase is instantly inactivated in contact with cytochrome *c*-552(s) (14 kDa) in vitro, it is protected from the inactivation by rusticyanin (Yamanaka et al., 1991a). Thus, the reduction of rusticyanin with ferrous ion catalyzed by the oxidoreductase in the presence of cytochrome *c*-552(s) (14 kDa) continues for an appreciable period in vitro. The reduced forms of both cytochrome *c*-552(s) (14 kDa) and rusticyanin are oxidized with oxygen by the catalysis of cytochrome *c* oxidase. On the basis of the results mentioned above, the electron transfer coupled to the oxidation of ferrous ion by *A. ferrooxidans* Fe1 (JCM 7811) will occur as shown in Fig. 5.1 (Yamanaka and Fukumori, 1995).

Although the reduced forms of cytochromes *c*-552(m) (22.3 kDa) (or cytochrome $c_4$) and *c*-550(m) (51 kDa) are also oxidized with molecular oxygen by the catalysis of cytochrome *c* oxidase, it has not yet been verified whether these cytochromes are reduced with ferrous ion by the catalysis of Fe(II)-cytochrome *c* oxidoreductase. On the basis of the studies of DNA that encodes the redox proteins of *A. ferrooxidans*, it is suggested that cytochrome $c_4$ is the direct electron donor for cytochrome *c* oxidase in vivo (Appia-Ayme et al., 1999). However, as already mentioned, the reactivity with cytochrome $c_4$ of cytochrome *c* oxidase is much lower than that of cytochrome *c*-552(s) (14 kDa) and is depressed with sulfate, while the enzymatic reaction of cytochrome *c*-552(s) (14 kDa) is stimulated by the salt. So cytochrome *c*-552(s) (14 kDa) seems to function as the real electron donor for the oxidase, if cytochrome *c* oxidase catalyzes the reduction of molecular oxygen at the outside of the plasma membrane as already mentioned (see also Fig. 5.2).

The redox proteins participating in the oxidation of ferrous ion other than cytochrome *c* oxidase are thought to be present in the periplasm, and they are fairly strongly bound to the outside of the plasma membrane. Thus, the spheroplasts of the bacterium retain 80% of the ferrous-oxidizing activity which the intact cells

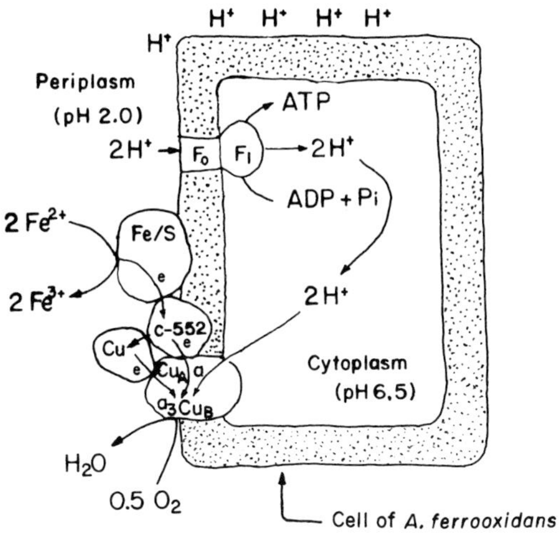


**Fig. 5.2.** A schematic presentation of the location of the electron transfer components participating in the oxidation of ferrous ion by an *Acidithiobacillus ferrooxidans* Fe1 (JCM 7811) cell. *Fe/S*, Fe(II)-cytochrome *c* oxidoreductase; *c-552*, cytochrome *c*-552(s) (14 kDa); *Cu*, rusticyanin; $C_Aaa_3Cu_B$, cytochrome *c* oxidase; $F_0F_1$, ATP synthase ($F_0F_1$-ATPase)

contain (Bodo and Lundgren, 1974). Therefore, the components participating in the oxidation of ferrous ion of the bacterium appear to be located in a manner shown schematically in Fig. 5.2.

As the optimal pH of cytochrome *c* oxidase of the bacterium is at pH 3.5 in vitro, the oxidase can catalyze the reduction of oxygen at the outside of cytoplasm membrane in vivo. As the pH difference between cytoplasm and periplasm is about 4.5, ATP will be synthesized by the catalysis of ATP synthase ($F_0F_1$-ATPase) when proton ($H^+$) in the periplasm is consumed through ATP synthase by the action of cytochrome *c* oxidase even if the membrane potential is zero. The ATP synthesis by utilizing proton gradients is verified with plasma membrane vesicles of *A. ferrooxidans* (Apel et al., 1980). The occurrence of ATP synthase in *A. ferrooxidans* is deduced from the finding of DNA for $F_1$ ATPase (Brown et al., 1994). Actually, some membrane potential will be generated because free energy of about 16 kcal/2$Fe^{2+}$ is liberated during the oxidation of ferrous ion with oxygen.

Yarzabal et al. (2002b) hypothesized that as *A. ferrooxidans* oxidizes pyrite, which is insoluble, the initial electron acceptor for electron from ferrous ion in its respiratory pathway should be located in the outer membrane, and that cytochrome *c* (46 kDa) occurring in the outer membrane is a candidate for the initial electron acceptor. They proposed a pathway of electron transfer from ferrous ion to oxygen as follows:

$$Fe^{2+} \rightarrow \text{Cyt } c \text{ (46 kDa)} \rightarrow \text{Rusticyanin} \rightarrow \text{Cyt } c \ (c_4\text{-type}) \\ \rightarrow \text{Cyt } c \text{ oxidase } (aa_3\text{-type}) \rightarrow O_2$$

However, Fe(II)-cytochrome *c* oxidoreductase may occur in the outer membrane, although it seems difficult to detect. Thus, it has been reported that an iron-lattice (polynuclear ferric complex) on the surface of *A. ferrooxidans* cells might function in the oxidation ferrous ion (Ingledew and Houston, 1986). Namely, it seems plausible that the polynuclear ferric complex is the same substance as Fe(II)-cytochrome *c* oxidoreductase which has many Fe/S clusters, and that the oxidoreductase oxidizes not only ferrous ion but also solid pyrite.

Takai et al. (2001) have obtained from *A. ferrooxidans* TI-1, a cytochrome *a*-type iron oxidase. The enzyme catalyzes the oxidation of ferrous ion at pH 3.0 and at 45°C in vitro; $V_{max}$ for oxygen consumption is 13.8 μmol $mg^{-1}$ $min^{-1}$ (242 mol $Fe^{2+}s^{-1}$). As the enzyme has neither cytochrome *c* nor cytochrome *b*, the ferrous ion-oxidizing pathway is constituted of only cytochrome *a* in the bacterium. Cytochrome *c* oxidase purified from a mercury-resistant strain of *A. ferrooxidans* was shown to reduce mercuric ion to elemental mercury (Sugio et al., 2001).

The iron-oxidizing bacteria need NAD(P)H for the reduction of carbon dioxide to produce the cellular materials. The electron transfer pathway in the reduction of $NAD^+$ with ferrous ion by *A. ferrooxidans* is proposed as follows (Elbehti et al., 2000).

$$Fe^{2+} \rightarrow c \rightarrow bc_1 \rightarrow QH_2/Q \rightarrow \text{NDH-1} \rightarrow NAD^+$$

where *c*, $bc_1$, $QH_2$, Q, and NDH-1 mean cytochrome *c*, cytochrome $bc_1$, quinol, quinone, and NADH dehydrogenase-1, respectively. The results have been obtained from the findings that the electron transfer from ferrocytochrome *c* to $NAD^+$ in the cell-free extracts is stimulated by inhibition of cytochrome *c* oxidase on addition of potassium cyanide. It is indicated also by other researchers (Brasseur et al., 2002) that cytochrome $bc_1$ complex functions in the reduction of $NAD(P)^+$ by *A. ferrooxidans*.

## 5.1.2 Oxidation of Sulfur Compounds by Iron-Oxidizing Bacteria

*Acidithiobacillus ferrooxidans* grows on reduced sulfur compounds as well as on ferrous ion and thus has oxidation processes of sulfur compounds. Sugio et al. (1987, 1989, 1991, 1992b) reported that most of the oxidation reactions of sulfur compounds in the bacterium proceed by the participation of ferric ion. Elemental sulfur is nonenzymatically reduced to hydrogen sulfide with reduced type of glutathione (GSH) [reaction (5.1)]. Hydrogen sulfide formed is oxidized with ferric ion by the catalysis of sulfide-Fe(III) oxidoreductase and sulfite is formed [reaction (5.2)]. The resulting sulfite is oxidized to sulfate by ferric ion by the catalysis of sulfite-Fe(III) oxidoreductase [reaction 5.3)]. In these reactions ferric ion is reduced to ferrous ion, and ferrous ion thus formed is oxidized through the iron-oxidizing system [reaction (5.4)].

$$S^{\circ} + 2GSH \xrightarrow{\text{Nonenzymatically}} H_2S + GSSG \tag{5.1}$$

$$H_2S + 6Fe^{3+} + 3H_2O \xrightarrow{\text{Sulfide-Fe(III) oxidoreductase}} SO_3^{2-} + 6Fe^{2+} + 8H^+ \tag{5.2}$$

$$SO_3{}^{2-} + H_2O + 2Fe^{3+} \xrightarrow{\text{Sulfite-Fe(III) oxidoreductase}} SO_4{}^{2-} + 2H^+ + 2Fe^{2+} \quad (5.3)$$

$$8Fe^{2+} + 8H^+ + 2O_2 \xrightarrow{\text{Iron oxdizing system}} 8Fe^{3+} + 4H_2O \quad (5.4)$$

However, it seems very peculiar that optimal pHs of both sulfide-Fe(III) oxidoreductase and sulfite-Fe(III) oxidoreductase are 5–7, although these enzymes occur in the periplasm (Sugio et al., 1987).

Some arguments are made against the oxidation mechanisms of sulfur compounds described above. Namely, although the oxidation of elemental sulfur with ferric ion in the bacterium is sensitive to *n*-heptylhydroxyquinoline-*N*-oxide (HOQNO), the reactions (5.1)–(5.3) which occur in the periplasm cannot explain the inhibitory effect of HOQNO on the oxidation of elemental sulfur with ferric ion (Corbett and Ingledew, 1987). Indeed, the oxidation of sulfite with ferric ion in the bacterium is not inhibited by HOQNO (Sugio et al., 1992b). Furthermore, the molar growth yield of the bacterium for reduced sulfur compounds is larger than that for ferrous ion (Hazeu et al., 1987). As the enzymes participating in the oxidation processes of sulfur compounds in (5.1)–(5.3) occur in the periplasm (Sugio et al., 1987), the difference in the growth yield mentioned above is not explained when electrons from the sulfur compounds enter into the electron transfer system through ferrous ion (Pronk et al., 1991a).

The inhibitory effects on the oxidation of ferrous ion and elemental sulfur differ between azide and cyanide (Harahuc et al., 2000). This seems to suggest also that the oxidation of ferrous ion and elemental sulfur are oxidized in different respective pathways, especially catalyzed by different oxidases. Harahuc and Suzuki (2001) claim that the oxidation of sulfite by *A. ferrooxidans* at pH 3 is catalyzed by free radicals, such as like $SO_3^{\bullet-}$, $SO_4^{\bullet-}$, and $OOSO_3^{\bullet-}$.

### *5.1.3 Various Growth Aspects of* **Acidithiobacillus Ferrooxidans**

*Acidithiobacillus ferrooxidans* grows under various conditions as well as it grows aerobically on ferrous ion and on reduced sulfur compounds. The bacterium grows electrophoretically by oxidation of electrons supplied from the cathode through the reaction of $Fe^{3+} \rightarrow Fe^{2+}$ (Kinsel and Umbreit, 1964; Taya et al., 1991; Blake et al., 1994). The autotrophic bacterium which grows on electrons supplied by electrolysis is called an electrotroph (Nakasono et al., 1997; Matsumoto et al., 1999). Furthermore, Ohmura et al. (2002) have shown that the bacterium grows anaerobically utilizing the electrode reactions [reactions (5.5)–(5.7)].

$$2H^+ + 2e \rightarrow H_2 \quad \text{(at cathode)} \quad (5.5)$$

$$H_2 + 2Fe^{3+} \rightarrow 2H^+ + 2Fe^{2+} \quad \text{(by bacterial cells)} \quad (5.6)$$

$$2Fe^{2+} - 2e \rightarrow Fe^{3+} \quad \text{(at anode)} \quad (5.7)$$

*Acidithiobacillus ferrooxidans* grows on formate at concentrations below 100 μM but its growth is inhibited by the compound at higher concentrations than 100 μM

(Pronk et al., 1991b). The molar growth yield for formate of the bacterium ($Y_{formate}$) is 1.32 g (dry weight), while that for ferrous ion ($Y_{Fe^{2+}}$) is 0.23 g (dry weight). The oxidation system of formate is inhibited by HOQNO (Corbett and Ingledew, 1987).

The growth of the bacterium is inhibited by benzoic acid, sorbate, and sodium laurylate (Onysko et al., 1984), and nitrate at 50 mM inhibits completely the oxidation of ferrous ion by the bacterium (Eccleston et al., 1985). Although the bacterium is sensitive to chloride ion, it becomes resistant to 140 μM chloride ion by training (Shiratori and Sonta, 1993). The bacterium is fairly resistant to heavy metal ions; its activity to oxidize ferrous ion is scarcely inhibited in the presence of 65 mM cupric ion, 100 mM nickel ion, 100 mM cobalt ion, 100 mM zinc ion, 100 mM cadmium ion, and 0.1 mM silver ion (Eccleston et al., 1985). The bacterium acquires the ability to grow even in the presence of 2 mM uranyl ion (Martin et al., 1983). Furthermore, it becomes resistant to arsenate and arsenite by training; a strain of the bacterium has been obtained which oxidizes ferrous ion in the presence of 80 mM arsenite and 287 mM arsenate (Collinet and Morin, 1990; Leduc and Ferroni, 1994). The resistant ability of the bacterium to arsenite and arsenate is important when they are applied for the solubilization of arsenopyrite (FeAsS) [reactions (5.8) and (5.9)]. *Leptospirillum ferrooxidans* is generally more sensitive to heavy metal ions than *A. ferrooxidans* (Eccleston et al., 1985).

$$2FeAsS + 5.5O_2 + 3H_2O \xrightarrow{\text{A. ferrooxidans}} 2H_3AsO_3 + 2FeSO_4 \tag{5.8}$$

$$2FeAsS + 6.5O_2 + 3H_2O \xrightarrow{\text{A. ferrooxidans}} 2H_3AsO_4 + 2FeSO_4 \tag{5.9}$$

*Acidithiobacillus ferrooxidans* anaerobically oxidizes elemental sulfur and formate with ferric ion (Sugio et al., 1985; Pronk et al., 1991a; Das et al., 1992). The bacterial oxidation of elemental sulfur and formate with ferric ion is inhibited by HOQNO but not by azide. In the oxidation of elemental sulfur with ferric ion free energy of 75 kcal/S° is liberated, which is enough to support the growth of the bacterium. Thus, the bacterium grows anaerobically by oxidizing elemental sulfur with ferric ion (Pronk et al., 1992).

An iron-oxidizing bacterium has been found that lacks some enzymes in the carbon dioxide-fixing system but can grow by oxidizing ferrous ion at pH 2.2 and at 48°C if the inorganic medium is supplemented with yeast extract or a mixture of six amino acids (glutamic acid, aspartic acid, alanine, arginine, histidine, serine) (Sugio et al., 1995). Moreover, it reduces ferric ion in the presence of the six amino acids. Other iron-oxidizing bacteria have also been found that grow by oxidizing ferrous ion at 50°C if the inorganic medium containing ferrous ion is supplemented with yeast extract (Brierley and Lockwood, 1977). These bacteria may also lack some enzymes in the carbon dioxide-fixing system.

Bacteria are also known that oxidize manganese (II). Although *Leptothrix discophora* was previously thought to be an iron-oxidizing bacterium which grew at neutral pH, it has now been found that this bacterium does not oxidize ferrous ion but rather manganese (II) ion ($Mn^{2+}$) (Mulder, 1989). Besides this bacterium, *Citrobacter freundii* oxidizes manganese (II) ion to manganese (IV) oxide (or manganese dioxide), although these bacteria are not lithoautotrophs (Ehrlich, 1984; Ehrlich et al., 1991). However, *Pseudomonas* S-36 grows on an inorganic medium containing manganese (II) ion if supplemented with a mixture of vitamins (Kepkay and Nealson, 1987).

### *5.1.4 Iron-Oxidizing Bacteria Requiring No Oxygen*

It is generally accepted that molecular oxygen did not occur in the biosphere of the primordial Earth. The age in which molecular oxygen appeared on Earth is thought to be recognized based on the age of the geological strata containing iron (III) oxide. However, because bacteria have been found that anaerobically oxidize ferrous ion to ferric ion, it is now suggested that the formation of iron (III) oxide is not necessarily related to the occurrence of molecular oxygen.

One of the bacteria classified into the purple bacteria group has been known to oxidize photosynthetically ferrous carbonate and ferrous sulfide to ferric hydroxide, resulting in the production of organic compounds [reactions (5.10) and (5.11)] (Ehrenreich and Widdel, 1994). Then ferric hydroxide is changed to iron (III) oxide.

$$4FeCO_3 + 10H_2O \xrightarrow{\text{light}} 4Fe(OH)_3 + [CH_2O] + 3HCO_3^- + 3H^+ \quad (5.10)$$

$$4FeS + 9HCO_3^- + 10H_2O + H^+ \xrightarrow{\text{light}} 4Fe(OH)_3 + 4SO_4^{2-} + 9[H_2O] \quad (5.11)$$

$[CH_2O]$: organic compounds biosynthesized. $Fe(OH)_3$ should be more exactly described as $FeO(OH) \cdot H_2O$.

*Chlorobium ferrooxidans*, a green sulfur phototrophic bacterium, also oxidizes ferrous carbonate anaerobically to ferric hydroxide utilizing light energy, and produces organic compounds and carbon dioxide if a *Geospirillum*-like bacterium co-resides [reaction (5.12)] (Heising et al., 1999).

$$17FeCO_3 + 29H_2O \xrightarrow{\text{light}} 17Fe(OH)_3 + [C_4H_7O_3] + 13CO_2 \quad (5.12)$$

$[C_4H_7O_3]$: organic compound biosynthesized

Furthermore, a bacterium has been found that oxidizes ferrous compounds to ferric hydroxide with nitrate and produces nitrite, hydrogen carbonate ion, and proton [reaction (5.13)]. The bacterium is called *Ferroglobus placidus*, an archaeon growing at 65–95°C (Hafenbradl et al., 1996). This bacterium thus produces ferric hydroxide without molecular oxygen or light.

$$2FeCO_3 + NO_3^- + 6H_2O \rightarrow NO_2^- + 2Fe(OH)_3 + 2HCO_3^- + 2H^+ + H_2O \quad (5.13)$$

*Dechlorosoma suillum* oxidizes ferrous compounds in their ores to ferric compounds with nitrate without dissolving the compounds (Chaudhuri et al., 2001). Moreover, magnetite is formed by the reaction of Fe(II) with Fe(III) which is formed during the oxidation of Fe(II) by the action of *D. suillum* using nitrate as the oxidant. The bacterium contributes to producing ores containing ferric compounds. It has been considered until now that a stratum containing iron (III) oxide confirms that cyanobacteria grew and produced molecular oxygen in the period when the stratum was formed. However, the presence of the bacteria mentioned above that oxidize ferrous compounds to ferric compounds without molecular oxygen suggests that iron (III) oxide (formed easily from ferric hydroxide) could have been produced without molecular oxygen. Therefore, the presence of iron (III) oxide in a stratum dating from an ancient era is not necessarily a straightforward proof for the presence of molecular oxygen in the biosphere of Earth at the time of the stratum containing iron (III) oxide, i.e., neither is it proof of the presence of cyanobacteria on Earth at the time of the stratum.

### *5.1.5 Bacterial Reduction of Ferric Compounds*

In the preceding section, the bacteria that oxidize ferrous compounds to ferric compounds have been described. In this section, those bacteria that reduce ferric compounds to ferrous compounds or oxidize organic or inorganic compounds with ferric compounds will be described. The bacteria of the *Geobacter* genus are strict anaerobes that oxidize hydrogen gas and some organic compounds with ferric ion. *Geobacter metallireducens* (Lovley et al., 1993) oxidizes several short chain fatty acids, alcohols, and monoaromatic compounds with ferric ion. As a result, ferric ion is reduced to ferrous ion. Furthermore, the bacterium oxidizes acetate with manganese (IV), uranium (VI), and nitrate. The bacterium is able to reduce iron (III) in insoluble iron (III) oxide without solubilization of the oxide (Nevin and Lovley, 2000). It is observed with the whole cell suspension of the bacterium that a *c*-type cytochrome in the organism is oxidized by the physiological acceptors, i.e., iron (III).

A membrane-bound NADH-dependent ferric iron reductase has been obtained from *Geobacter sulfurreducens* (Magnuson et al., 2000). The enzyme contains a hemoprotein and FAD. the reduced hemoprotein in the enzyme is reoxidized on addition of ferric ion and NADH is a specific electron donor for the enzyme. The bacterium has a citric acid cycle (TCA cycle) (Galushko and Schink, 2000). Besides *G. metallireducens* and *G. sulfurreducens*, several bacteria of *Geobacter* genus have been isolated, namely *G. bremensis*, *G. pelophilus* (Straub and Buchholz-Cleven, 2001), *G. hydrogenophilus*, *G. chapelli*, and *G. grbiciae* (Coates et al., 2001).

Around the hydrothermal vents in the deep sea, extremely thermophilic bacteria reside which have the optimal growth temperature at 100°C which participate in the reduction of several metals. For example, *Pyrobaculum islandicum* (Kashefi and Lovley, 2000) and *Geoglobus ahangari* (Kashefi et al., 2002) reduce ferric iron with hydrogen gas at 100°C and around 90°C, respectively. When poorly crystalline ferric oxide is reduced by *P. islandicum*, magnetite ($Fe_3O_4$) is formed extracellularly during its reduction. *Pyrobaculum islandicum* also reduces metals other than iron (III) with hydrogen gas as the electron donor; e.g., it reduces uranium (VI) to insoluble uranium (IV) and reduces manganese (IV) to manganese (II). Manganese (II) once formed makes manganese carbonate (Kashefi and Lovley, 2000). The resulting compounds produce ores and the ores formed are deposited around the hydrothermal vents. From the vents, uranium (VI) compounds come forth which are water soluble, and are reduced by *P. islandicum* to uranium (IV) compounds which are in turn insoluble and are deposited around the vents to make uranium ores (e.g., uraninite). Furthermore, the bacterium is able to reduce gold (III) ion to metal gold with hydrogen gas (Kashefi et al., 2001). The compounds, metals, or ores thus formed are deposited around the hydrothermal vents. If the vents are heaved up to create a mountain after 10–100 million years, a mine seam is produced in which the ores mentioned above are involved. Usually, gold is found in the elemental or metal state. This will be explainable in some cases if gold has been formed by the action of the iron-reducing bacteria as mentioned above.

### 5.1.6 Bacteria Containing Magnetism

Although magnetotactic bacteria are not lithoautotrophs, they are included here in relation to the oxidation and reduction of iron. In wetland marshes, the bacteria that have magnets, i.e., magnetotactic bacteria, reside (Blakemore, 1982). One of the most well known magnetotactic bacteria is *Magnetospirillum magnetotacticum*. In its cells, many particles of 40–100 μm diameter are present making up a line. Each particle is called a magnetosome, which is a crystal of magnetite ($Fe_3O_4$) covered with phospholipid membrane (Fukumori, 2000). A magnetosome is a magnet, and a bacterium can recognize the magnetic field by the magnetosome. The bacteria residing in the Northern Hemisphere access the south pole of the magnet and escape from the north pole, while those that reside in the Southern Hemisphere access the north pole of the magnet and escape from the south pole. Thus, a bacterium residing in the Northern Hemisphere moves toward the North Pole (of Earth), while that residing in the Southern Hemisphere moves toward the South Pole. Why do the bacteria behave in such a manner? As the bacterium grows in environments where the partial pressure of oxygen is considerably low, it will move in a northern direction, in the case of the Northern Hemisphere, probably to move toward a place of low oxygen pressure. The processes by which the magnetite is formed in the bacterial cells have not been clarified. Magnetite seems to be formed during the enzymatic oxidation of ferrous ion to ferric ion in the bacterial cells, and cytochrome $cd_1$-type nitrite reductase seems to participate in its formation (Yamazaki et al., 1995).

## 5.2 Applications of Iron-Oxidizing Bacteria

### *5.2.1 Bacterial Leaching*

Copper is easily extracted with 5% sulfuric acid from the ores containing cupric hydroxide, cupric carbonate, and cupric oxide such as azurite, malachite, and tenorite. However, copper is seldom extracted with sulfuric acid from copper ores containing cupric sulfides; the presence of ferric ion is mandatory to extract copper with sulfuric acid from the ores, e.g., chalcopyrite [$FeCuS_2$].

$$FeCuS_2 + 2Fe_2(SO_4)_3 + 2H_2O + 3O_2 \rightarrow CuSO_4 + 5FeSO_4 + 2H_2SO_4 \quad (5.14)$$

*A. ferrooxidans*

After copper has been extracted as cupric sulfate from chalcopyrite, ferric ion added is reduced to ferrous ion, and sulfuric acid is formed [reaction (5.14)], and the reaction solution becomes very acidic, around pH 2. Although it is necessary for successive leaching copper to reoxidize ferrous ion to ferric ion, ferrous ion is not oxidized spontaneously with atmospheric oxygen even with a vigorous aeration as the pH of the reaction solution is around 2. If the acidophilic iron-oxidizing bacteria are present, ferrous ion is easily oxidized to ferric ion and the leaching of copper proceeds continuously. The processes in which metals are leached by the action of bacteria, as seen above, are called bacterial leaching (Lundgren and Silver, 1980), or in a wider sense, bioleaching (Bosecker, 1997). The illustration shown in Fig. 5.3 shows a schematic presentation of an apparatus used for the bacterial leaching of, e.g., chalcopyrite on a laboratory scale. First, the solution of ferric sulfate is showered over the ores containing chalcopyrite, and the acidic solution containing cupric sulfate leached out from the ores flows away. Secondly, metal copper is precipitated from the leached solution on addition of metal iron. Thirdly, the resulting ferrous sulfate is reoxidized to ferric sulfate by aeration in the presence of *Acidithiobacillus ferrooxidans*. Then the ferric sulfate solution is pumped up to be sprayed again over the ores. Thus, copper in the ore is continuously leached.

When the laboratory processes are applied to a strip copper mine, a long plastic pipe with many small holes is spread over the mine and the solution of ferric sulfate is sprayed on the ores. The solution containing cupric sulfate flows down the mine hill, collects in a pool made of acid-resistant concrete, and crude metal copper is obtained from the solution by substitution with metal iron. Besides copper, molybdenum, bismuth, zinc, etc. are leached from the respective ores using the bacterium. However, at present bacterial leaching is not carried out in Japan.

At this point the author would like to refer to the collection of uranium using bacteria. In Japan, bacterial leaching of uranium has been tried experimentally

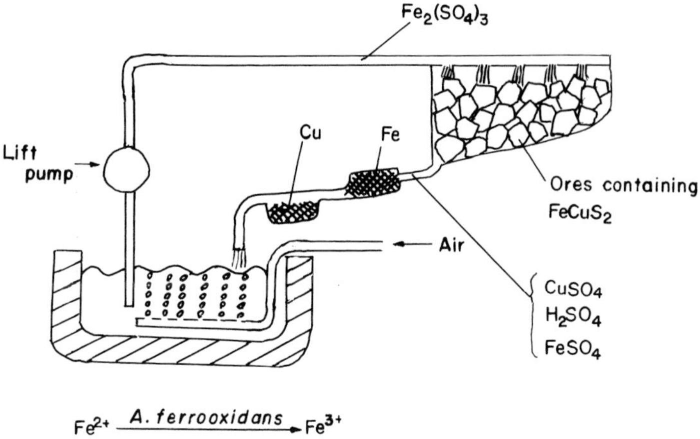


**Fig. 5.3.** A diagram of the apparatus for the bacterial leaching on a laboratory scale. The acidophilic iron-oxidizing bacteria (e.g., *Acidithiobacillus ferrooxidans*) are present in the pool and oxidize ferrous ion to ferric ion

(Tomizuka and Takahara, 1972). The uranium ore produced in Japan is Ningyo-rock [$CaU(PO_4)_2$]. As the compound contains calcium and phosphate other than uranium, uranium is not successfully leached.

$$CaU(PO_4)_2 + Fe_2(SO_4)_3 + H_2SO_4 + 2H_2O \rightarrow UO_2SO_4 + 2FeSO_4 + 2H_3PO_4 + CaSO_4 \quad (5.15)$$

(*A. ferrooxidans*: $FeSO_4 \rightarrow Fe_2(SO_4)_3$)

On the contrary, in Canada and other countries, pitchblende [$UO_{2+x}$] is produced, which reacts with ferric sulfate to form only soluble uranium compound, uranyl sulfate [$UO_2SO_4$]. As a result, ferric sulfate is changed to ferrous sulfate [reaction (5.16)]. When ferrous sulfate is reoxidized to ferric sulfate by use of *A. ferrooxidans*, uranium is continuously leached from uranium oxide as uranyl sulfate. In Canada, previously the bacterium was sprayed with 9K medium (culture medium for the acidophilic iron-oxidizing bacteria; Silverman and Lundgren, 1959) on the ground of the abandoned mine, and 50 tons of uranium (as $U_3O_8$) per year was recovered (MacGregor, 1964; Fisher, 1966). Nowadays the studies on the leaching of uranium are continuing using the method described above in Spain and Brazil (Cerdá et al., 1993).

(*A. ferrooxidans*: $FeSO_4 \rightarrow Fe_2(SO_4)_3$)

$$UO_2 + Fe_2(SO_4)_3 \rightarrow UO_2SO_4 + 2FeSO_4 \quad (5.16)$$

(In the reaction formula, the composition of pitchblende is simplified as $UO_2$). Notice that $UO_2$ in which uranium is of valency 4 is not soluble in water, while $UO_2SO_4$ containing valency 6 uranium is soluble.

### 5.2.2 *Etching of Copper Plate*

The dissolution of copper with ferric sulfate is not limited to the leaching of copper from copper ores, but ferric ion also dissolves metal copper; ferric ion is used to produce copper etching and to make electronic circuitry. However, the waste solution used for etching is not recycled. During etching, ferric chloride but not ferric sulfate is used. The acidophilic iron-oxidizing bacteria are sensitive to chloride ion. Even in the presence of at most 0.0008% (47 μM) ferric chloride, the bacteria do not oxidize ferrous ion (cf. p. 90), while they grow on ferrous ion in the presence of 5% (ca. 123 000 μM) ferric sulfate. If the etching of copper is performed by use of ferric sulfate, the waste may be recycled using the bacteria. Why is ferric sulfate not used for the etching in place of ferric chloride? The power of ferric sulfate to dissolve copper is much weaker than that of ferric chloride; it takes too long for the sulfate to dissolve copper as compared to the chloride. As the two compounds dissociate to form ferric ion, they appear to be equally effective in the etching process. So why does the difference occur between the two compounds? Probably, the difference depends on the degree of dissociation of the salts. Thus, ferric chloride is easily soluble in absolute ethanol, while ferric sulfate is scarcely soluble in the alcohol (see, e.g., *Encyclopedic Dictionary of Chemistry*, Tokyo-Kagakudojin, Japan, 2001).

### 5.2.3 *Concentration of Gold from Pyrite Containing a Trace of Gold*

In Mexico and South Africa, pyrite containing traces of gold and silver [$FeS_2$(Au,Ag)] is produced (Claassen et al., 1993; Inoue et al., 1995). The pyrite ore contains 8 g gold and 43 g silver in 1 000 000 g. Therefore, even if sodium cyanide (NaCN; dissolving reagent for gold and silver) is applied to the pyrite the reagent does not reach the metals. When the pyrite ore is treated with the acidophilic iron-oxidizing bacteria, sulfur and iron in the ore are dissolved, while gold and silver remain in a solid state; the content of gold and silver is considerably condensed. Although the colors of the metals are not yet seen at this stage, gold and silver are dissolved on addition of sodium cyanide as sodium dicyanoaurate (I) [$NaAu(CN)_2$] and sodium dicyanoargentate (I) [$NaAg(CN)_2$], respectively. Then metallic gold and metallic silver are obtained by reduction of the salts with metallic zinc. (The reaction formulae are presented in the following only for gold).

$$FeS_2(Au) + 3.5O_2 + H_2O \rightarrow Fe^{2+} + 2SO_4^{2-} + Au + 2H^+ \quad (5.17)$$

$$2Au + 4NaCN + 2H_2O \rightarrow 2NaAu(CN)_2 + 2NaOH + H_2 \quad (5.18)$$

$$2NaAu(CN)_2 + Zn \rightarrow Na_2Zn(CN)_4 + 2Au \quad (5.19)$$

### *5.2.4 Biohydrometallurgy*

As mentioned above, at present bacterial leaching is not performed in Japan. However, the acidophilic iron-oxidizing bacteria are applied to biohydrometallurgy.

In the Kosaka mine of Akita Prefecture, Japan, Kuroko (black-ore) is produced. Kuroko is an assemblage of ores obtained from the mine which is thought to have been formed by the heaving of the hydrothermal vents containing deposits of salts of various metals (mainly sulfides) in the deep sea. Kuroko contains various metals; the average metal contents per 1 ton of 7 Kuroko ores are: copper (Cu) 22 kg, lead (Pb) 160 kg, iron (Fe) 53 kg, zinc (Zn) 225 kg, arsenic (As) 1030 g, gold (Au) 3 g, and silver (Ag) 312 g (Ishikawa, 1991). To refine metals from Kuroko, the ores are first heated at a high temperature (in a flashsmelting furnace) to melt them. Crude copper mass resulting from cooling of the molten metals contains gold and silver. On further refining of crude copper by electrolysis, gold and silver are obtained as the anode slimes. On the other hand, the smoke ash (flue dust) obtained when the ores are heated at high temperature contain lead, copper, iron, zinc, and arsenic. To refine these elements, the ash is dissolved in diluted sulfuric acid, and thereafter the process comprises hydrometallurgy. When the ash is dissolved in diluted sulfuric acid, lead is precipitated as sulfate. After the precipitate of lead sulfate is recovered, copper dissolved in the supernatant as cupric sulfate is precipitated as sulfide by blowing hydrogen sulfide into the solution. The remaining zinc, iron, and arsenic occur as zinc ion ($Zn^{2+}$), ferrous ion ($Fe^{2+}$), and arsenate ion ($AsO_4^{3-}$), respectively. Next, ferrous ion is oxidized to ferric ion by the action of the acidophilic iron-oxidizing bacteria. Ferric ion formed is combined with arsenate ion to precipitate as ferric arsenate. When hydrogen sulfide and ammonia are added to the resulting supernatant, zinc ion is precipitated as zinc hydroxide (Fig. 5.4). Although previously much electric power was required to oxidize ferrous ion by heating, the application of bacteria to oxidize ferrous ion has greatly reduced the consumption of electricity.

### *5.2.5 Cleaning of Mine Sewage*

In Iwate Prefecture, Japan there is an abandoned mine near the Akagawa river, upstream of the Kitakamigawa river. The Matsuo mine was abandoned in 1971 and since then a great amount of mine sewage has been leaching from the mine. The sewage contains sulfuric acid and large amounts of ferrous ion. Previously, while

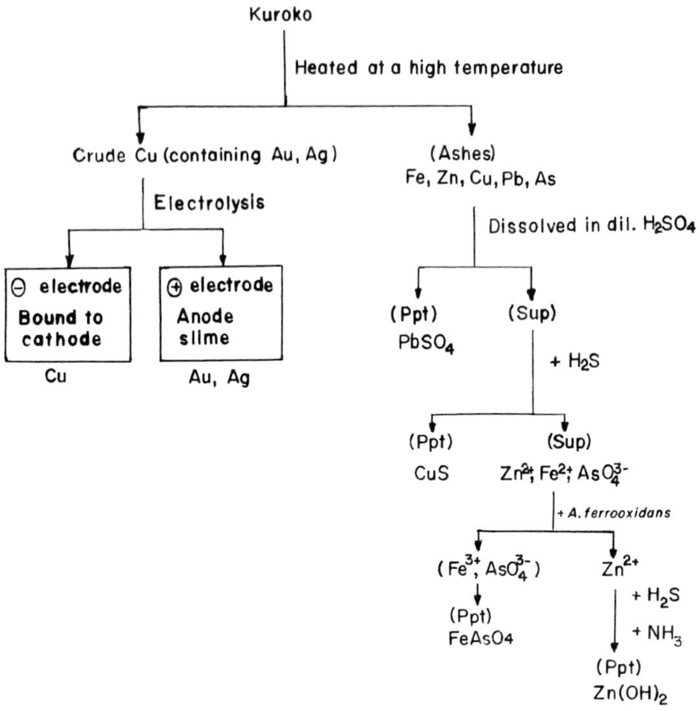


**Fig. 5.4.** A scheme briefly presenting the refining processes of metals from Kuroko. Although the real manufacturing processes are much more complicated, the scheme insists that zinc ion is separable from ferric arsenate after ferrous ion has been oxidized by the action of the acidophilic iron-oxidizing bacterium (or bacteria)

the sewage flowed down the Akagawa river to the lower parts, ferrous ion was oxidized to ferric ion and the resulting ferric hydroxide was precipitated on the bottom of the Akagawa river, creating reddish brown slime. To protect the river from the ferric hydroxide damage, previously the mine sewage was neutralized with calcium hydroxide (lime) and calcium carbonate before the sewage reached the river. Although clean water was obtained, i.e., the water contained no iron ions, and its pH was raised to about neutral, the by-products formed by these technical processes, ferric hydroxide and calcium sulfate, made a mixed colloidal precipitate; the by-products were not only unutilizable but also unmanageable. When ferrous ion in the sewage is oxidized to ferric ion by the action of the acidophilic iron-oxidizing bacteria before the neutralization, ferric ion is precipitated as ferric hydroxide by raising the pH of the sewage to 3–4 by addition of comparatively smaller amounts of calcium carbonate, but calcium sulfate is not precipitated at this stage. When further calcium carbonate is added to the resulting supernatant to raise the pH to neutral, calcium sulfate is precipitated. The supernatant obtained is clean and dischargeable, and the two kinds of the precipitates separately obtained are utilizable (Imai, 1984). Thus the Akagawa has been restored now to a normal, clean river.

## 5.3 Upheaval of House Foundations: Damage Caused by Bacteria

In Iwaki City, Fukushima Prefecture, Japan, about 1000 wooden houses have been damaged by heterogeneous upheaval of house foundations; pillars are tilted and walls are broken by the heaving of the ground under floors of houses. The damage by the upheaval of the foundations is seen also in factory buildings; the floor paved with concrete is heaved heterogeneously and supports are necessary to keep the equipment level. The maximal height of upheaval known up to now is 48 cm. Most of the houses damaged by the upheaval are built directly on nonweathered mudstone of the Neogene. The ground under the house built in the area where the heaving occurs sometimes smells of hydrogen sulfide at the onset of the damage. The pH of the ground is lowered to about 3 from 7–8 when the ground begins to heave, and crystals of gypsum and jarosite appear. Then damages such as like tilting of pillars occurs (Yohta, 2000).

In the ground where the upheaval occurs, sulfate-reducing, sulfur-oxidizing, and acidophilic iron-oxidizing bacteria reside. When the mixture of mudstone in the area and the medium for the sulfate-reducing bacteria (Postgate, 1979) is anaerobically incubated at 37°C, hydrogen sulfide is formed. When the mudstone is treated at 121°C for 20 min, the formation of hydrogen sulfide is not observed. After 10 days of incubation, the supernatant obtained by sedimenting the mudstone particles contains the bacteria as shown in Fig. 5.5.

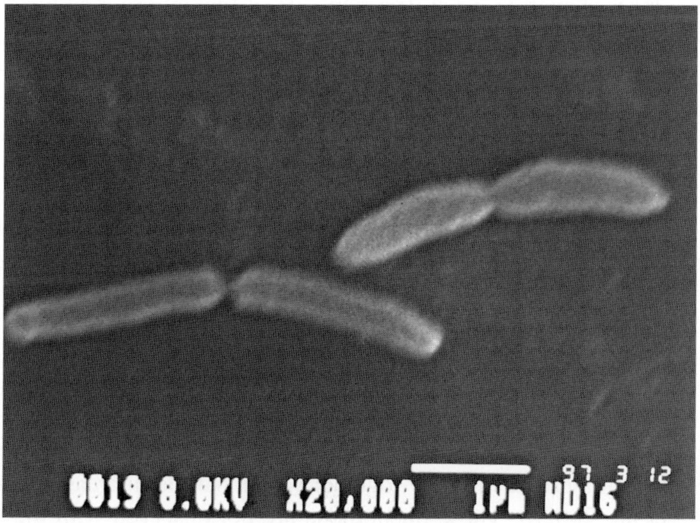


**Fig. 5.5.** Scanning electron micrograph, gold-shadowed, of the bacterial cells obtained from the 10 days culture when the mixture of mudstone and the medium for the sulfate-reducing bacteria was anaerobically incubated at 37°C. *Bar*: 1 μm

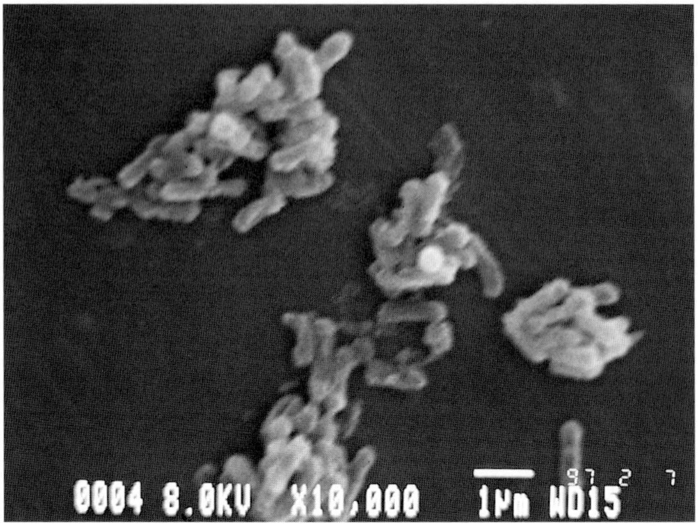


**Fig. 5.6.** Scanning electron micrograph, gold-shadowed, of the bacterial cells obtained from the 7 days culture when the mixture of the mudstone and 9K medium of Silverman and Lundgren (1959) was shaken in air at 30°C. *Bar*: 1 μm

When the mixture of the mudstone and the medium for the sulfur-oxidizing bacteria (pH 6.5) [e.g. S6 medium, American Type Culture Collection Catalogue (1982), or inorganic medium of Santer et al. (1959) supplemented with 0.2% sodium sulfide] is shaken in air at 28°C, the pH of the culture medium is lowered from 6.5 to about 2. The lowering of the pH is not observed with the mudstone treated at 121°C for 20 min. The results show that the sulfur-oxidizing bacteria are present in the mudstone. When the mixture of the mudstone and 9K medium (pH 2.0, containing 50 g ferrous sulfate per liter) (Silverman and Lundgren, 1959) is shaken in air at 30°C, the amount of ferric ion increases. The increase in the amount of ferric ion is not observed when the mudstone is treated at 121°C for 20 min. The results show that acidophilic iron-oxidizing bacteria reside in the mudstone. Thus, the supernatant obtained by sedimenting the soil particles from the 7 days of culture contain the bacteria as shown in Fig. 5.6.

From the shape (rods) of the bacteria, they appear to be *Acidithiobacillus ferrooxidans* but not *Leptospirillum ferrooxidans*. Finally, when the mixture of the mudstone and the medium for the sulfur-oxidizing bacteria (pH 6.5) supplemented with powdered pyrite is shaken in air, the pH of the culture medium is lowered over time. When the pH is lowered to below 4, the amount of ferrous ion plus ferric ion in the medium increases rapidly together with a parallel increase of sulfate ion (Yamanaka et al., 2002b). These phenomena are not observed with the mudstone heated at 121°C for 20 min. The results show that the acidophilic iron-oxidizing bacteria (growing at pH lower than about 4) oxidize pyrite but the usual sulfur-

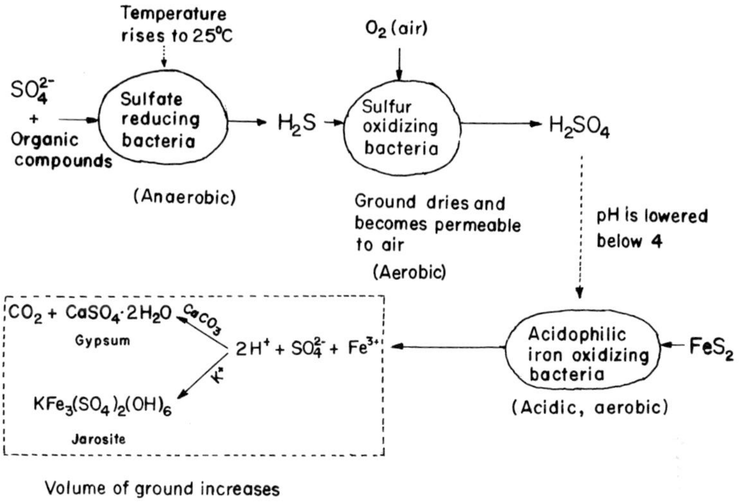


**Fig. 5.7.** A schematic presentation for the principle of the ground upheaval caused by bacteria

oxidizing bacteria do not oxidize pyrite, because while the pH of the culture medium is lowered to approximately 4, iron ions do not appear. The amount of sulfate increased is twice as much as the amount of the sum of ferrous and ferric ion increase. The oxidation of pyrite by the acidophilic iron-oxidizing bacteria proceeds as shown by the reaction formulas (5.20) and (5.21).

$$2FeS_2 + 7O_2 + 2H_2O \rightarrow 2Fe^{2+} + 2H_2SO_4 + 2SO_4{}^{2-} \qquad (5.20)$$

$$4Fe^{2+} + O_2 + 4H^+ \rightarrow 4Fe^{3+} + 2H_2O \qquad (5.21)$$

The experimental results obtained above show also that pyrite is not easily oxidized chemically even if it is in contact with air and water for several days, although it may be oxidized when it is exposed to similar conditions for several years or more.

The mechanisms by which the upheaval of the foundations occurs will be considered as follows. As the nonweathered mudstone contains much water (52%–53%) (Oyama et al., 1998), the environment on the inside of the mudstone is anaerobic. The place where the sulfate-reducing bacteria is present moves near to the surface of the ground by the excavation for leveling, while the temperature rises from 17°C to 25°C in the summertime. As the content of organic compounds is fairly high in the mudstone (6%–7%) (Oyama et al., 1998), the sulfate-reducing bacteria grow actively as the temperature rises and produce hydrogen sulfide. As the ground under the house dries out gradually because of, e.g., shielding of rains and lowering of the ground water level, the ground becomes permeable to air. The sulfur-oxidizing bacteria start growing and oxidize hydrogen sulfide to produce sulfuric acid. As a result, the pH of the ground is lowered to below 4, then the acidophilic iron-oxidizing bacteria grow actively and oxidize pyrite to produce much sulfuric acid as the mudstone contains 5% pyrite (Oyama et al., 1998). In the oxidation of pyrite, first ferrous ion is formed and the ferrous ion is further oxidized to ferric ion. Sulfuric acid formed reacts with calcium carbonate in the ground to form calcium sulfate (gypsum) and carbon dioxide, and reacts with potassium ion and ferric ion to produce jarosite. By the formation of crystals of these minerals and carbon dioxide gas, the ground is bulked up. This is the cause of the subsequent upheaval of the house foundations (Yamanaka et al., 2002b) (Fig. 5.7).

# Chapter 6
# Carbon Circulation on Earth and Microorganisms

At present, carbon occurs as carbon dioxide in the atmosphere in the ratio of 0.033% of air. Green plants (including algae and cyanobacteria) convert carbon dioxide to organic compounds by photosynthesis and evolve molecular oxygen. Animals eat the compounds and convert the compounds again to carbon dioxide. Thus carbon is circulated on Earth (Fig. 6.1).

When the so-called fossil fuels such as coal and petroleum are used a large amount of carbon dioxide is released, resulting in an increase of carbon dioxide in the atmosphere. As carbon dioxide is a gas with an influence on the greenhouse effect, its increase in the atmosphere results in the raising of the temperature on the surface of Earth. Therefore the increase of carbon dioxide in the atmosphere should be curtailed. It is usually thought that most carbon dioxide in the atmosphere is consumed by plants on the land. Of course this idea is not incorrect; the tropical rain forests of the Amazonian area, Kalimantan (Borneo) and some other areas consume much of the carbon dioxide in the atmosphere. However, it should be kept in mind that the consumption of the gas by algae and cyanobacteria in the ocean is also sizable. Some researchers say that 70% of carbon dioxide in the atmosphere is consumed by the photosynthesis of algae and cyanobacteria in the oceans.

Although plants, algae, and cyanobacteria photosynthetically consume carbon dioxide and evolve oxygen in the daytime, they respire at night; in the dark they not only cease evolving oxygen but they also evolve carbon dioxide. In the 1770s when J. Priestley discovered oxygen, people in Europe began to put flowers in vases and potted plants in rooms to cleanse the air. The practice has continued of taking the flowers and plants outside at night after people became aware that the plants evolve carbon dioxide in the dark. However, the amount of carbon dioxide which plants, algae, and cyanobacteria evolve during the night is far lower than that which they absorb in the daytime.

As described in the preceding chapters, chemolithoautotrophic bacteria consume carbon dioxide regardless of light, both in the daytime and at night. They need carbon dioxide to form cellular materials and oxidize inorganic compounds to acquire energy without evolving carbon dioxide. It appears to be possible, therefore, that these bacteria reduce a great amount of carbon dioxide. Unfortunately, the

*Chemolithoautotrophic Bacteria*. T. Yamanaka
doi: 10.1007/978-4-431-78541-5_6, © Springer 2008

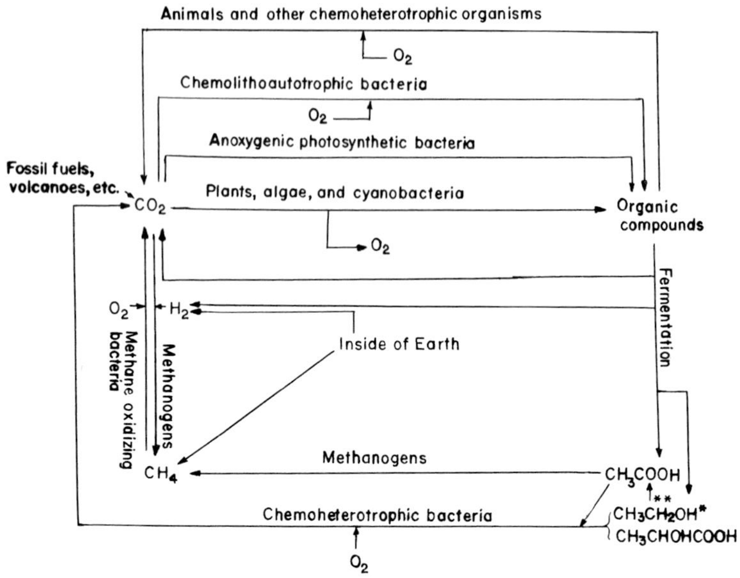


**Fig. 6.1.** A summary of natural circulation of carbon. *Asterisk* indicates that other alcohols and fatty acids are also formed; *double asterisk*, anaerobic and aerobic oxidations

number of cells of these bacteria is extremely small compared with that of the cells of algae and cyanobacteria, so that the consumption of carbon dioxide by the chemolithoautotrophic bacteria is far lower than that by the algae and cyanobacteria.

Sugars and starch biosynthesized by plants are oxidized by animals and other chemoheterotrophic organisms and carbon dioxide is formed. Furthermore, organic compounds are fermented to organic compounds of lower molecular weights (in most cases lactic acid, acetic acid, and ethanol) and, in many cases, carbon dioxide. In some cases, hydrogen gas is evolved during the fermentation processes. Lactic acid, acetic acid, and ethanol produced are also changed to carbon dioxide by the chemoheterotrophic bacteria, and acetic acid is changed to methane by the methanogens.

Although the crust of the Earth and the oceans are large reservoirs of carbon dioxide gas, physical circulation of the gas between these parts and the biosphere is not considered in this book. However, some amount of the carbon dioxide dissolved in the oceans will be circulated through algae and cyanobacteria.

In many cases, a minute amount of carbon monoxide is formed when organic compounds are burned. Even if the amount of the gas is minute, it cannot be ignored as it is very toxic. Some of the bacteria form organic compounds from carbon monoxide. In this chapter, the mechanisms by which bacteria form organic compounds from carbon monoxide will be described.

## 6.1 Mechanisms of Formation of Organic Compounds from Carbon Dioxide

The lithoautotrophs have to form cellular materials from carbon dioxide. The process to change carbon dioxide into organic compounds is called fixation of carbon dioxide. On the basis of the knowledge to date, all algae and cyanobacteria, and many of the plants, fix carbon dioxide through the Calvin–Benson cycle (or reductive pentose phosphate cycle) (Bassham et al., 1954), while the plants of 20 families and 1200 species have been known to fix carbon dioxide through the Hatch–Slack pathway (or $C_4$ dicarboxylate pathway) (Hatch et al., 1967).

Many chemolithoautotrophic bacteria biosynthesize organic compounds by fixing carbon dioxide through the Calvin–Benson cycle (Aleem, 1970). However, methanogens, obligately autotrophic hydrogen-oxidizing bacteria, and some anoxygenic photosynthetic bacteria fix carbon dioxide by pathways other than the Calvin–Benson cycle.

### *6.1.1 Calvin–Benson Cycle (Reductive Pentose Phosphate Cycle)*

Plants, algae, cyanobacteria, and chemolithoautotrophic bacteria have to form cellular materials from carbon dioxide. How is this done? Many autotrophic organisms catch (fix) carbon dioxide by the catalysis of the enzyme called Rubisco (ribulose-1,5-bisphosphate carboxylase/oxygenase) (see Fig. 6.2). The fixation entails the binding of carbon dioxide with ribulose-1,5-bisphosphate to form 3-phosphoglycerate. Next, 3-phosphoglycerate formed is reduced to glyceraldehyde-3-phosphate by the successive catalyses of phosphoglycerate kinase and glyceraldehyde-3-phosphate reductase using ATP, and NADPH or NADH, respectively. The formation of NADPH or NADH requires energy; light energy is utilized in the case of the photosynthetic organisms, while the energy is supplied by the oxidation of inorganic compounds in the case of the chemolithoautotrophs. Half of the glyceraldehyde-3-phosphate formed is changed to dihydroxyacetone phosphate, and this reacts with glyceraldehyde-3-phosphate to form fructose-1,6-bisphosphate. Fructose-1,6-bisphosphate is changed to glucose-6-phosphate after releasing one phosphate. Next, glucose-6-phosphate is changed to glucose-1-phosphate, and this compound is further changed to UDP-glucose (uridine 5′-diphosphate-glucose) and ADP-glucose (adenosine 5′-diphosphate-glucose). UDP-glucose and ADP-glucose are used for the biosyntheses of sucrose and starch, respectively. For the further continuous fixation of carbon dioxide, a continuous supply of ribulose-1,5-bisphosphate is required. This compound is formed using part of glyceraldehyde-3-phosphate; glyceraldehyde-3-phosphate forms ribulose-5-phosphate via many processes (Fig. 6.2), and ribulose-5-phosphate is changed to ribulose-1,5-bisphosphate by the catalysis of phosphopentokinase with ATP. The energy for formation of ATP is derived from the light energy in the case of the photosynthetic organisms, while it

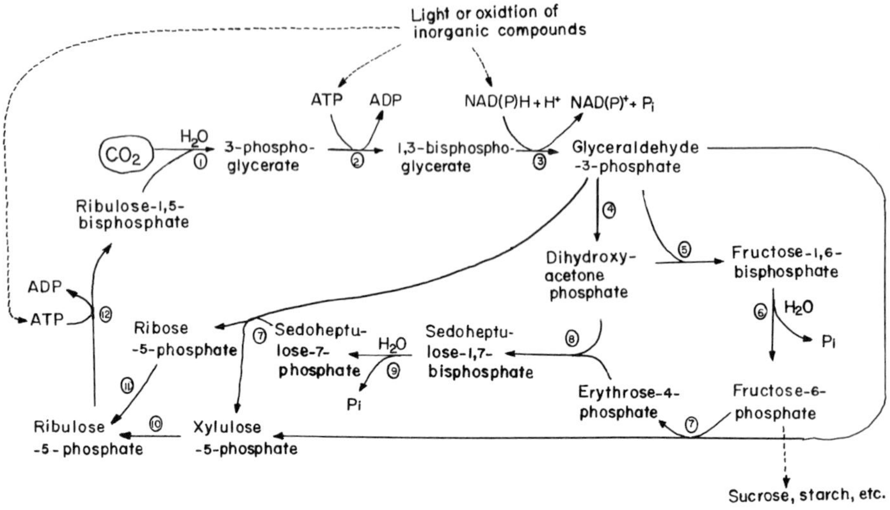


**Fig. 6.2.** A revised form of the Calvin–Benson cycle (prepared by a modification of Bassham et al., 1954). In the original Calvin–Benson cycle, only NADPH is used for the reduction of 3-phosphoglycerate as the reductant and the only energy source for the formation of ATP is light. In the cycle functioning in the anoxygenic photosynthetic bacteria and in the chemolithoautotrophs, NADPH and NADH [NAD(P)H] are used as the reductants. Moreover, NAD(P)H and ATP are formed using energy liberated from the oxidation of inorganic compounds without light in the case of the chemolithoautotrophs. Thus, the cycle shown in this figure is named as a revised form of the cycle. *Circled numbers*: *1*, ribulose-1,5-bisphosphate carboxylase/oxygenase (Rubisco); *2*, phosphoglycerate kinase; *3*, glyceraldehyde-3-phosphate dehydrogenase; *4*, triose-phosphate isomerase; *5*, fructose-bisphosphate aldolase; *6*, fructose-bisphosphatase; *7*, transketolase; *8*, aldolase; *9*, sedoheptulose-bisphosphatase; *10*, ribulose-phosphate 3-epimerase; *11*, ribose-5-phosphate isomerase; *12*, phosphoribulokinase. Pi, phosphate

is supplied from the oxidation of inorganic compounds in the case of chemolithoautotrophs. The reactions for the formation of ribulose-1,5-bisphosphate thus mentioned make a cycle together with the reactions for carbon dioxide fixation (Fig. 6.2). This cycle is known as the Calvin–Benson cycle (Bassham et al., 1954). The cycle was established mainly by the cooperation of Drs. Melvin Calvin and Andrew Benson. However, the Nobel Prize was awarded to only Dr. Calvin, so in many cases the cycle is called the Calvin cycle. However, researchers of photosynthesis who know that the cycle was established through the cooperation of the two persons call it the Calvin–Benson cycle, as it is described in this book.

### *6.1.2 Hatch–Slack Pathway*

Plants are divided into two groups for the fixing mechanisms of carbon dioxide; plants of one group use the Calvin–Benson cycle, while those of the other group use pathways other than the Calvin–Benson cycle. For example, sugar cane fixes carbon dioxide by a reaction with phosphoenolpyruvate (PEP) but not with ribulose-1,5-biphosphate.

$$CH_2 = \underset{\displaystyle OPO_3H_2}{\underset{|}{C}}\text{-}COOH + CO_2 + H_2O \xrightarrow{\text{PEP carboxylase}} \underset{\displaystyle CH_2COOH}{\underset{|}{C}}OCOOH + H_3PO_4 \qquad (6.1)$$

As a result, oxaloacetate (OAA, $C_4$-compound) is formed unlike the case of the Calvin–Benson cycle in which 3-phosphoglycerate ($C_3$-compound) is formed. The pathway in the fixation of carbon dioxide by the catalysis of PEP-carboxylase is observed in sugar cane, corn, etc., and is called the Hatch–Slack pathway (Hatch et al., 1967). The plants having the Hatch–Slack pathway have chloroplasts both in mesophyll cells and in vascular bundle sheath cells, and the Hatch–Slack pathway occurs in the mesophyll cells. Oxaloacetate formed by the fixation of carbon dioxide in the mesophyll cells is reduced to malate. Malate thus formed moves to the vascular bundle sheath cells and releases carbon dioxide there. Carbon dioxide released is fixed by the catalysis of Rubisco, and the organic compounds are formed through the Calvin–Benson cycle. (Fig. 6.3).

The plants producing organic compounds from carbon dioxide through the Calvin–Benson cycle are called $C_3$-plants, while the plants producing organic compounds from carbon dioxide through the Hatch–Slack pathway are called $C_4$-plants.

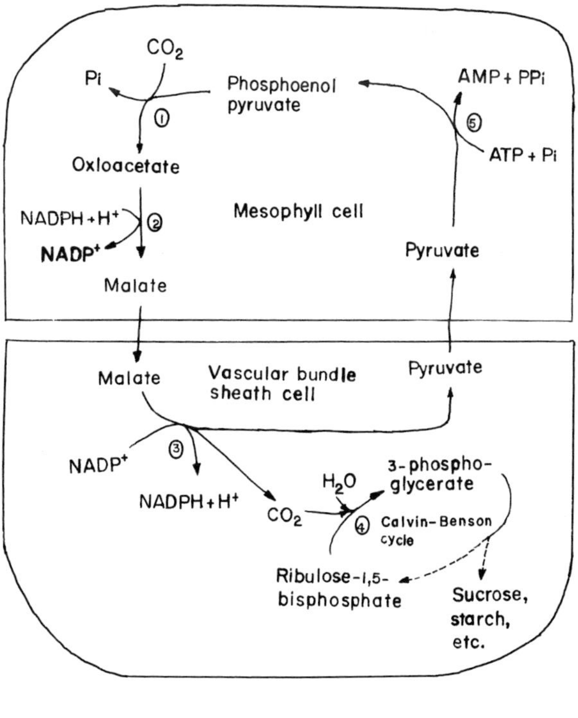


**Fig. 6.3.** Carbon dioxide-fixing pathway in the $C_4$-plants (prepared mainly on the basis of Hatch et al., 1967). *Circled numbers*: *1*, phosphoenolpyruvate carboxylase; *2*, malate dehydrogenase ($NADP^+$); *3*, malate dehydrogenase ($NADP^+$) (OAA decarboxylating) (= "malic" enzyme); *4*, Rubisco; *5*, pyruvate orthophosphate dikinase. Pi, phosphate; PPi, diphosphate

**Table 6.1.** Comparison between $C_3$- and $C_4$-plants

| | $C_3$-plants | $C_4$-plants |
|---|---|---|
| Examples of plants | Spinach, soybean, rice plant | Sugar cane, corn, Japanese millet |
| Optimal temperature for photosynthesis (°C) | 10–25 | 30–40 |
| Amount of $CO_2$ fixed [(mg/100 cm$^2$ leaf) per h] | 15–40 | 40–80 |
| Effect of light intensity on photosynthetic rate | Saturated at 1–2 × $10^4$ lux | Proportional limitlessly to the intensity of natural light |
| Water use efficiency | Low | High |
| Effect of $CO_2$ on photosynthetic rate | Half of maximal rate at 0.033% $CO_2$ | Saturated at conc. of $CO_2$ lower than the case of $C_3$-plant |
| Photorespiration (activity to decompose organic compounds and evolve $CO_2$ during illumination) | Strong | Extremely weak |

Some of the properties of the $C_3$-plants are compared with those of the $C_4$-plants in Table 6.1. As the $C_4$-plants fix carbon dioxide by the catalysis of phosphoenol-pyruvate carboxylase but not by that of Rubisco, several differences occur between the $C_3$-plants and $C_4$-plants. The $C_4$-plants have higher activity in producing organic compounds and higher efficiency in utilization of light energy and water than the $C_3$-plants. Rice is a $C_3$-plant, while Japanese millet is a $C_4$-plant. In summertime we sometimes see that much more Japanese millets than rice plants grow in the rice paddies. Japanese millets utilize the light energy of the sun much more efficiently than rice plants do, and grow faster than rice plants. For Japanese people living on rice it seems unfortunate that rice plants are not $C_4$-plants. However, as the $C_4$-plants have their optimal growth temperature at 30–40°C, they do not grow optimally in the Temperate Zone. As the chemolithoautotrophic bacteria and cyanobacteria do not have the Hatch–Slack pathway, the organisms corresponding to the $C_4$-plants do not occur in these bacteria.

### *6.1.3 Carbon Dioxide-Fixing Pathways Other than the Calvin–Benson Cycle in the Lithoautotrophs*

As already mentioned, cyanobacteria and most of the chemolithoautotrophic bacteria fix carbon dioxide through the Calvin–Benson cycle, but some litho-autotrophic bacteria fix carbon dioxide through other pathways. When the green phototrophic bacterium *Chloroflexus aurantiacus* grows lithoautotrophically, the bacterium fixes carbon dioxide through the 3-hydroxypropionate cycle (Ivanovsky et al., 1993; Strauss and Fuchs, 1993). The cycle seems to function in fixing carbon dioxide also in an archaeon, *Acidianus brierleyi* (Ishii et al., 1997). In the chemolithoautotrophic hydrogen-oxidizing bacterium *Hydrogenobacter thermophilus*, a reductive carboxylate cycle functions in the fixation of carbon dioxide (Shiba et al., 1985). The reductive carboxylate cycle was first found in the phototrophic green sulfur bacteria (Evans et al., 1966; Fuchs et al., 1980a,b). In the hydrogen-oxidizing bacterium, ATP is synthesized using the energy liberated by the oxidation of hydrogen gas with molecular oxygen, and NADH and reduced ferredoxin as the reducing agents for carbon dioxide are formed utilizing hydrogen gas (Yoon et al., 1997), unlike in the green sulfur bacteria which form NAD(P)H and reduced ferredoxin using light energy. The biosynthetic pathway of organic compounds in the methanogens has not been completely clarified. The pathway seems to be a combination of acetyl-CoA synthesizing system with the reductive carboxylate cycle in *Methanobacterium thermoautotrophicum* and *Methanosarcina barkeri* (Fuchs and Stupperich, 1978; Bhatnagar et al., 1991).

The acetogenic bacteria produce one molecule of acetate from two molecules of carbon dioxide, and the carbon dioxide-fixing reaction in the bacteria proceeds through acetyl-CoA synthesizing pathways. A carbon monoxide dehydrogenase (CODH) plays an important role in the acetyl-CoA synthesizing pathway. The enzyme contains 6 nickel atoms, 3 zinc atoms, 32 nonheme iron atoms, and 42 inorganic sulfides in a 440 kDa molecule (Ragsdale et al., 1983a,b). The CODH molecule has three binding sites, respectively for methyl-, carbon monoxide, and -SCoA, and synthesizes acetyl-CoA on the enzyme molecule. Methyl group is biosynthesized from carbon dioxide through many reaction processes involving tetrahydrofolic acid. Then a corrinoid enzyme participates in the transfer of methyl group to CODH (Pezacka and Wood, 1986; Wood and Ljungdahl, 1991).

## 6.2 Methanogens

Methane is a gas which has a greenhouse effect more than 20 times that of carbon dioxide. Therefore, the generation of this gas should be curtailed as positively as possible. For example, methane is released from the bottom of a dirty swamp. In this case, methanogens are present at the bottom and they produce methane by oxidizing hydrogen with carbon dioxide. All of the methanogens form methane by oxidizing hydrogen with carbon dioxide, while some of them form the gas by reducing $C_1$-compounds like methanol, methylamines, formate and carbon monoxide, and acetate in addition to the reduction of carbon dioxide. In any case, methanogens biosynthesize ATP coupling to the production of methane and grow using the energy derived from ATP (Blaut and Gottschalk, 1984).

The methanogens that biosynthesize ATP only by oxidizing hydrogen with carbon dioxide belong to the chemolithoautotrophs, and they produce the cellular materials also from carbon dioxide. Namely, for the methanogens, carbon dioxide is not only the energy source but also the substance for the formation of the cellular materials. This relation between bacteria and carbon dioxide is unique.

Although some persons use the term methane bacteria, the term is ambiguous; the bacteria that form methane should be called methane-generating bacteria, i.e., methanogens, while the bacteria that oxidize methane should be called methane-oxidizing bacteria. Incidentally, the methanogens belong to archaebacteria whereas the methane-oxidizing bacteria belong to eubacteria.

### *6.2.1 Mechanism of Lithoautotrophic Methane Formation: Respiration but Not Fermentation*

Although it was known by the end of the nineteenth century that methanogens form methane, the reaction mechanism of the methane formation was elucidated only as late as the 1980s (Rouviére and Wolfe, 1988) (Fig. 6.4). In methane formation by methanogens from carbon dioxide and hydrogen gas, carbon dioxide first reacts with methanofuran (MF) to form formyl methanofuran (formyl MF). Next, formyl group is transferred to tetrahydromethanopterin ($H_4MP$) to form formyl-$H_4MP$. Formyl-$H_4MP$ formed is successively reduced to methyl-$H_4MP$ (Vaupel and Thauer, 1998), and methyl group of this compound is transferred to coenzyme M (CoM, or CoMSH where the important SH group is shown) to form methyl CoM ($CH_3SCoM$). Then, methyl CoM reacts with coenzyme B (CoBSH) by the catalysis of methyl CoM reductase and methane is formed (Thauer and Bonacker, 1994). In this reaction, methyl group is transferred to nickel (I) of coenzyme $F_{430}$ in the reductase from methyl-CoM before being reduced to methane. Next, heterodisulfide (CoMS-SCoB) formed as the result of methane formation is reduced by the catalysis of $H_2$:heterodisulfide oxidoreductase to reform CoMSH and CoBSH. Recently, reduced coenzyme $F_{420}$-dependent heterodisulfide reductase has also been purified from *Methanococcus voltae* (Brodersen et al., 1999).

Although a series of the reactions in the methane formation do not make a cycle as for compounds, as the formation of methane stimulates the reaction of carbon dioxide with methanofuran a series of the reactions seem to make a cycle in a sense of function. So the series of the reactions is called the $C_1$ cycle (Fig. 6.5).

Among the reactions in the $C_1$ cycle mentioned above, those that liberate enough free energy to form ATP are those catalyzed by methyl CoM reductase [reaction (6.2)] and $H_2$:heterodisulfide oxidoreductase [reaction (6.3)] (Heiden et al., 1993). In the cycle, no organophosphate compounds are found, which can be the intermediate for ATP biosynthesis as observed in the fermentation.

$$CH_3SCoM + HSCoB \xrightarrow{\text{Methyl CoM Reductase}} CH_4 + CoMS\text{-}SCoB + 10.7\,kcal \quad (6.2)$$

$$CoMS\text{-}SCoB + H_2 \xrightarrow{H_2:\text{ heterodisulfide oxidoreductase}} HSCoM + HSCoB + 9.5\,kcal \quad (6.3)$$

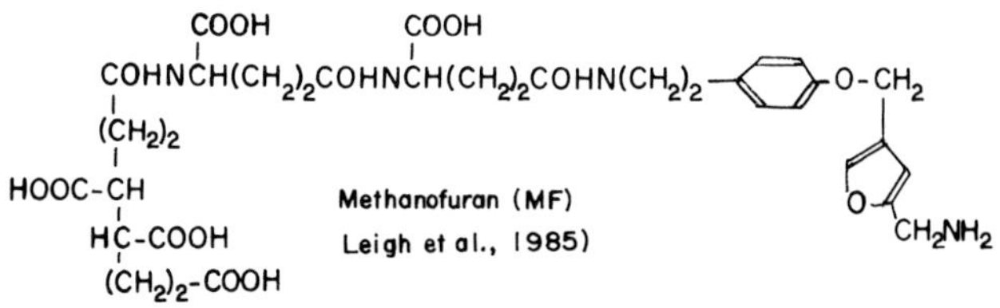

Methanofuran (MF)
Leigh et al., 1985)

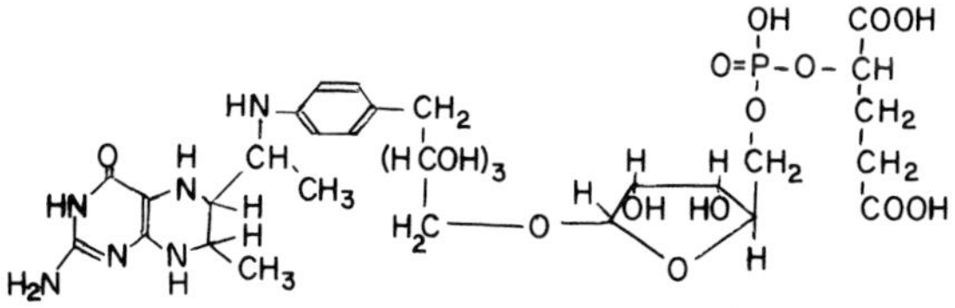

Tetrahydromethanopterin ($H_4MP$)
(Van Beelen et al., 1984)

$HSCH_2CH_2SO_3^-$

Coenzyme M (CoMSH)
(Taylor & Wolfe, 1974)

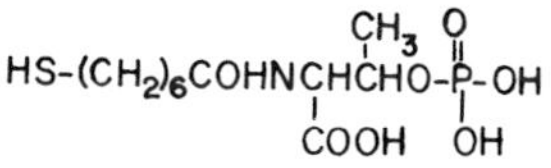

Coenzyme B (CoBSH or HS-HTP)
( Nell et al., 1986; Ellermann et al., 1988)

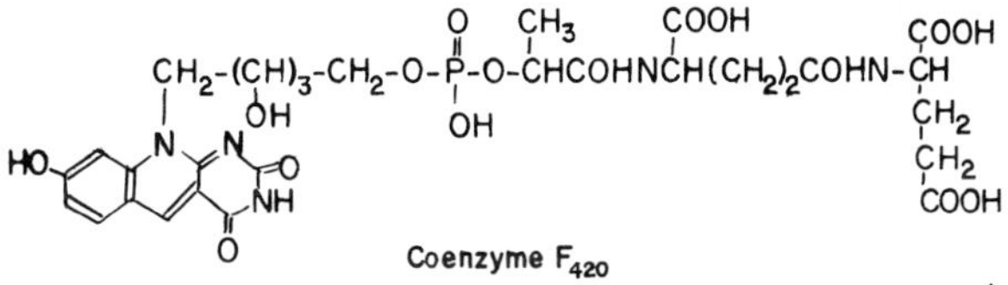

Coenzyme $F_{420}$
(Cheeseman et al., 1972; Eirich et al., 1978)

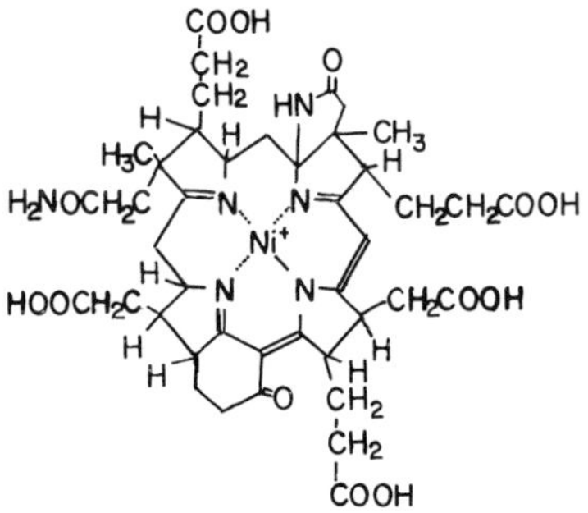

Coenzyme $F_{430}$
(Gunsalus & Wolfe, 1978; Diekert et al., 1980; Pfaltz et al., 1982)

**Fig. 6.4.** Structures of various coenzyme-like compounds characteristic of methanogens

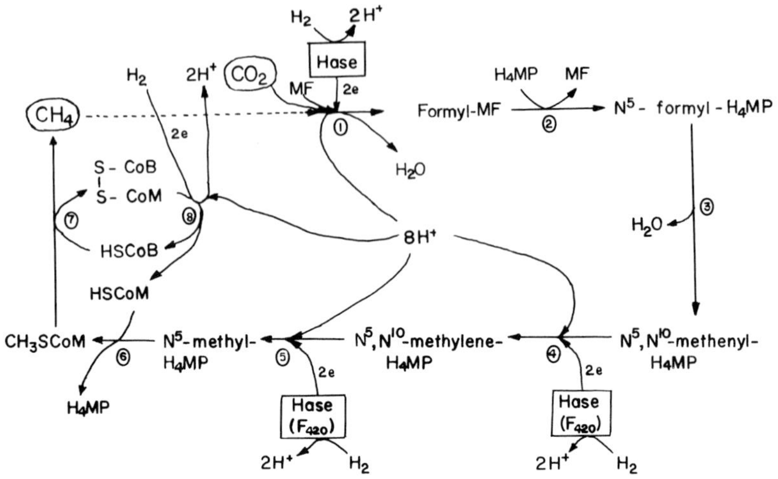


**Fig. 6.5.** A schematic presentation of the $C_1$ cycle (prepared on the basis of Rouviére and Wolfe, 1988; Thauer and Bonacker, 1994; Vaupel and Thauer, 1998). *Circled numbers*: *1*, formylmethanofuran dehydrogenase (contains Fe/S, and Mo, or W); *2*, formylmethanofuran-tetrahydromethanopterin *N*-formyltransferase; *3*, methenyltetrahydromethanopterin cyclohydrolase; *4*, methylenetetrahydromethanopterin dehydrogenase; *5*, methylenetetrahydromethanopterin reductase; *6*, tetrahydromethanopterin *S*-methyltransferase (contains Fe/S and vitamin $B_{12}$); *7*, methyl-CoM reductase (contains $F_{430}$); *8*, $H_2$:heterodisulfide oxidoreductase

Furthermore, the formation of ATP coupled to methane production by the methanogens is inhibited in the presence of ionophores (Blaut and Gottschalk, 1985). Indeed, an ATPase which seems to function as ATP synthase has been known to occur in the methanogens (Inatomi, 1986; Mayer et al., 1987). Thus, the formation of methane by the methanogens is a respiration but not fermentation; it is now called carbon dioxide respiration (or carbonic respiration) (Wolfe, 1980; Daniels et al., 1984).

When methane is formed from methanol ($CH_3OH$) and methyl amine ($CH_3NH_2$) by the action of methanogens, the methyl group of these compounds is used for the formation of methyl CoM, and methane is formed from this compound (Shapiro and Wolfe, 1980). Many methanogens form methane also from carbon monoxide. When the methanogens form methane from carbon monoxide, carbon monoxide is reduced at several reaction steps to form methyl-$H_4FA$. This compound is used to form acetyl-CoA. Then, methane is formed from acetyl-CoA through probably the same processes as seen in the methane formation from acetate (see below) (Stupperich et al., 1983).

The source materials for methanogens to form methane are limited to one carbon ($C_1$) compound except for acetate. Previously, *Methanobacterium omelianskii* was thought to form methane from ethanol (Barker, 1956). But the "bacterium" was a mixed culture of *Methanobacterium bryantii* and a bacterium which decomposed ethanol to form hydrogen gas (Bacterium S) (Bryant et al., 1967). Therefore, the name *Methanobacterium omelianskii* has now been deleted. Since then bacteria have been found which form methane from ethanol (Widdel, 1986). However, even in this case methane is not produced from ethanol itself, but ethanol donates hydrogen atoms to reduce carbon dioxide; $NADP^+$ is reduced with ethanol by the catalysis of alcohol dehydrogenase and the resulting NADPH is used through coenzyme $F_{420}$ for the reduction of methenyl-$H_4$MP and methylene-$H_4$MP (Berk and Thauer, 1997).

Methyl CoM reductase has coenzyme $F_{430}$ containing nickel as the prosthetic group (Gunsalus and Wolfe, 1978; Diekert et al., 1980; Pfaltz et al., 1982; Thauer and Bonacker, 1994). Therefore, nickel is necessary to the formation of methane by the methanogens. Coenzyme $F_{430}$ is the complex of a porphyrin with nickel. Although the structure of the porphyrin seems very different from those of the porphyrins for hemes as described already (p. 12), these porphyrins all are biosynthesized through the same pathway until the intermediate, uroporphyrinogen III (see Fig. 4.4, p. 60). The metalloporphyrins known in the organisms up to now are chlorophyll (Mg and Zn complexes), heme (Fe complex), vitamin $B_{12}$ (Co complex), and coenzyme $F_{430}$ (Ni complex). Nature seems to have given different functions to a compound, porphyrin nucleus by using different metals.

## *6.2.2 Formation of Methane from Acetate*

Many methanogens like *Methanosarcina* sp. and *Methanothrix thermophila* ALS-1 produce methane from acetate (Ferry, 1992).

$$CH_3COO^- + H^+ \rightarrow CH_4 + CO_2 \qquad \Delta G^{\circ'} = -8.6\,\text{kcal/mol} \tag{6.4}$$

This reaction releases the free energy enough for the biosynthesis of ATP.

In the formation of methane from acetate, acetate is first changed to acetyl CoA and then methyl CoM is formed from this compound [reactions (6.5)–(6.7)] (Ragsdale et al., 1983a,b; Grahame and Stadtman, 1987; Fischer and Thauer, 1988; Harder et al., 1989; Allen and Zinder, 1996).

$$CH_3COOH \xrightarrow{\text{Acetate kinase}} CH_3COOPO_3H_2 + ADP \tag{6.5}$$

$$CH_3COOPO_3H_2 + CoASH \xrightarrow{\text{Phosphate acetyltransferase}} CH_3COSCoA + H_3PO_4 \tag{6.6}$$

$$CH_3COSCoA + CoMSH + H_2O \xrightarrow{\text{CODH-corrinoid enzyme complex}} CH_3SCoM + HSCoA + CO_2 + 2H^+ + 2e \tag{6.7}$$

[CODH: carbon monoxide dehydrogenase]

Finally, methane is formed from methyl CoM by the catalysis of methyl CoM reductase (Ferry, 1992). When acetyl group ($CH_3CO$–) is decomposed to methyl group ($CH_3$–) and carbon monoxide on the CODH complex molecule, carbon monoxide reacts with water, and electrons are released. The electrons are used for the reduction of $H_2$: heterodisulfide oxidoreductase. A carbon monoxide:heterodisulfide oxidoreductase system has also been found; carbon monoxide reduces heterodisulfide oxidoreductase through the CODH complex, ferredoxin, and cytochrome *b* (Peer et al., 1994).

### 6.2.3 Methanogens and Cytochromes

Before 1950, it was generally accepted that cytochromes participate only in the (aerobic) respiratory system but they do not occur in the fermentation processes (Keilin, 1966). Therefore, the finding was very surprising in that the sulfate-reducing bacteria possessed cytochrome (cytochrome $c_3$) (Ishimoto et al., 1954; Postgate, 1954), because the bacteria were previously thought to acquire life energy by fermentation. Meanwhile, the reduction of sulfate by the sulfate-reducing bacteria was found to be respiration (sulfate respiration) and the participation of cytochrome in the bacterial reduction of sulfate has been understood to be rational. Furthermore, as the anaerobic photosynthetic bacteria *Chlorobium* and *Chromatium* were found to have cytochromes (Vernon and Kamen, 1954), the presence of cytochrome in the anaerobic bacteria has been widely accepted, and the participation of cytochrome in the photosynthetic electron transfer system has also been recognized. However, no researchers ever imagined that cytochrome was present in the methanogens until the 1970s. In 1979, Kühn et al. found cytochromes in the cells of *Methanosarcina* sp. Moreover, $H_2$:heterodisulfide oxidoreductase purified from *Methanosarcina barkeri* was found to have cytochrome *b* (Heiden et al., 1993) and cytochrome *bc* complex was purified from *M. barkeri* (Kumazawa et al., 1994). The reduced form of cytochrome *b* in the purified $H_2$:heterodisulfide oxidoreductase is oxidized rapidly on addition of heterodisulfide (Heiden et al., 1994). Furthermore, it has been found that in *Methanosarcina* sp., a cytochrome *b* is reduced with hydrogen by the catalysis of hydrogenase, ferrocytochrome *b* donates electrons to $H_2$:heterodisulfide oxidoreductase containing cytochrome *b* via methanophenazine, and this cytochrome *b* becomes a reduced form. Next, the reduced enzyme donates electrons to heterodisulfide and the ferrocytochrome *b* in the enzyme is oxidized (Hedderich et al., 1999). As the enzyme is important in the biosynthesis of methane, the presence of cytochrome *b* in the enzyme seems to show that cytochrome *b* is necessary to the formation of methane in all the methanogens. Thus, the methane formation in *M. barkeri* is stimulated by an enhanced biosynthesis of cytochrome by the increased ferrous ion concentration in the culture medium (Lin et al., 1990). However, no cytochromes have been found in the bacteria of *Methanobacterium* genus (Kühn and Gottschalk, 1983; Kühn et al., 1983). This means that cytochrome is not mandatory for the biosynthesis of methane by the methanogens. The findings that cytochromes are present in some of the methanogens and the methanogens grow by means of respiration will conceptualize more perfectly that cytochrome does participate in the respiration. The absence of cytochrome in

*Methanobacterium* may mean that the respiratory system has evolved from the primordial type having no cytochrome and the evolving steps are observed in *Methanobacterium*. Thus, $H_2$:heterodisulfide oxidoreductase of *Methanobacterium thermoautotrophicum* has been found to have FAD in place of cytochrome *b* of the *Methanosarcina* enzyme (Hedderich et al., 1999).

## 6.2.4 Methanogens and the Environment

The evolution of methane gas from the bottom of a dirty marsh means that the anaerobic bacteria are decomposing organic compounds at the bottom and the methanogens form methane from carbon dioxide and hydrogen formed by the other anaerobic bacteria there. This means a cleansing of the polluted marsh bottom is carried out by a cooperation of these bacteria. This phenomenon is applicable to treatment of waste water which contains so many organic compounds that it is difficult to make the water aerobic by even vigorous aeration. Methane gas produced from the digestion of the organic compounds by a cooperation of the anaerobic organisms with the methanogens is used as fuel after removing hydrogen sulfide; methane gas cleaned by removing hydrogen sulfide can be used as a fuel that releases 4000–6000 kcal/$m^3$ when it is burned.

As methane gas has a greenhouse effect more than 20 times that of carbon dioxide, the natural and artificial production of the gas should be curtailed as completely as possible. However, it is not easy to curtail the natural production of the gas. For example, in humans the methanogens reside in the intestines of one third of the population, and persons carrying the methanogens evolve the gas in exhalation and in wind from the anus (Koga, 1988).

The evolution of methane gas from rice paddies is a problem. The soil environments of the paddies are anaerobic as the ground of the paddies is covered by a water layer of approximately 3–5 cm in depth. In Japan, intermediate drainage is performed several times during midsummer to prevent the activity of the sulfate-reducing bacteria, as described above. The intermediate drying of the paddies is also known to prevent the formation of methane gas (Kimura et al., 1991). Therefore, in Japan farmers have been contributing to prevention of the production of methane by the paddies. Recently, an attempt has been reported which suppresses the production of methane by addition of a revolving furnace slag (a by-product of the steel industry) to the paddy soil (Furukawa and Inubushi, 2002). The ammonia-oxidizing bacteria oxidize methane as mentioned in the preceding section. If the bacteria and the methane-oxidizing bacteria reside in the oxidizing layer (surface layer) of the ground of the paddy field, both the bacteria will prevent the evolution of methane (Bédard and Knowles, 1989).

Much methane is included in the gas exhaled by ruminants such as cattle. In the rumen of the ruminant stomach, many anaerobic microorganisms reside and decompose food anaerobically to biosynthesize ATP. As the anaerobic organisms other than the methanogens decompose the organic compounds by releasing hydrogen from the compounds by the catalysis of hydrogenase, much hydrogen gas is formed. The enzyme catalyzes reversibly the evolution and absorption of hydrogen gas. So when the partial pressure of the gas becomes higher, the evolution of hydrogen stops. Then the anaerobes synthesizing ATP by evolution of hydrogen gas cannot live. The methanogens change hydrogen gas to methane which does not revert to hydrogen and carbon dioxide. Thus, the anaerobes including the methanogens grow actively and the organic compounds decomposed (and re-synthesized) become good food for the ruminants.

It was sometimes experienced in Europe that fasted cows became ill on eating too much feed rich in starch. When the organic compounds in the feed are decomposed slowly or normally, the evolution of hydrogen is also slow and the methanogens in the rumen can change hydrogen to methane. However, when hungry cows eat feed rich in starch, glycolysis will proceed very quickly and so much pyruvic acid and hydrogen atoms (bound to compound and/or hydrogen gas) are suddenly formed that the methanogens cannot treat it completely. Hydrogen should be treated by the reaction with pyruvic acid. Thus, a large amount of lactic acid is formed suddenly, and the acidity of the resulting lactic acid induces illness in the cow.

## 6.3 Bacteria Utilizing Carbon Monoxide

Although some bacteria grow on carbon monoxide (CO), little is known about the processes in utilizing carbon monoxide. Among the bacteria utilizing carbon monoxide there exist both anaerobic and aerobic bacteria. The aerobic bacteria have respiratory enzymes which catalyze the reduction of molecular oxygen. It seems very mysterious that although the enzymes appear to be inhibited by carbon monoxide, the aerobic bacteria grow in the atmosphere containing the gas, while cytochrome *c* oxidase and hemoglobin of humans and animals combine with the gas easily and as a result their respiration stops. The bacteria are useful not only for the synthesis of organic compounds from carbon monoxide but also for the cleansing of pollution by the gas. If many such bacteria grow actively in the soil at the side of roads, they will consume carbon monoxide in the exhaust from automobiles.

The bacteria utilizing carbon monoxide have a unique carbon monoxide dehydrogenase (CODH) containing nickel (Ragsdale et al., 1983a,b). On the surface of the enzyme molecule there are three sites that can bind three different substrates, respectively, and the three substrates bound to the enzyme react with each other on the enzyme molecule by the catalysis of the enzyme. For example, the synthesis of organic compounds from carbon monoxide in *Clostridium thermoaceticum* proceeds through the acetyl-CoA-synthesizing pathway. First, one molecule of carbon monoxide is changed to carbon dioxide by the catalysis of CODH. In this case, the enzyme functions as a "usual" dehydrogenase. The resulting carbon dioxide is successively reduced to methyl tetrahydrofolic acid ($H_4FA$). Then the methyl group in methyl-$H_4FA$ is transferred to a cobalt-containing enzyme (vitamin $B_{12}$-containing enzyme, corrinoid enzyme) (Hu et al., 1984; Ragsdale et al., 1987). Another molecule of carbon monoxide reacts with CODH to make CODH-CO and the resulting enzyme complex accepts the methyl group from methyl-corrinoid enzyme to form $CH_3$-CODH-CO. This enzyme complex reacts with CoA (or CoASH) to form acetyl-CoA ($CH_3COSCoA$). Acetyl-CoA is a good source for biosynthesis of various organic compounds.

# Chapter 7
# Organisms Evolutionarily Closest to the Origin of Life

It was reported in 1983 that bacteria were present around the hydrothermal vent at 2650 m in depth where the temperature was 350°C (Baross and Deming, 1983). Although it seemed curious that organisms could reside at 350°C, it was recognized at that time that the organisms could reside at a high temperature of 350°C, because the pressure is high at 2650 m depth (265 atm) and water remains in liquid form. Baross and Deming (1983) reported that they had succeeded in culturing the bacteria in their laboratory at 250°C and under 250 atm. The bacteria grew at the generation time of 40 min and methanogens were included among them. Namely, it was found that some of the methanogens could grow at 250°C and at 250 atm. From these findings it seemed that some of the methanogens might have resided even at the geological age when the temperature of the surface of Earth was still more than 100°C, even though their growing temperature, 250°C, was not exactly verified.

## 7.1 Archaea and Their Energy-Acquiring Reactions

Woese and Fox (1977) claimed that the methanogens constitute the third kingdom of the organisms based on the comparison of the base sequence of the oligonucleotides derived from 16S rRNA (18 S rRNA in the case of eukaryotes) and called the methanogens archaebacteria or archaea; the world of the organisms is composed of eukaryotes or eukarya, eubacteria or bacteria, and archaebacteria or archaea. They thought that the methanogens which grow anaerobically on a simple energy source such as carbon dioxide and hydrogen must be very primitive organisms, so they called the methanogens archaebacteria. Although since then the archaebacteria have been found not to be the oldest organisms evolutionarily, it is true that the archaebacteria constitute the third kingdom of the organisms. Namely, it has been found that, e.g., the phospholipids of the cell membranes are unique to the archaebacteria; the basal structure of phospholipids in the cell membranes of eukaryotes and eubacteria is diester of *sn*-glycerol-3-phosphate, while that of phospholipids in the cell membranes of archaebacteria is diether of *sn*-glycerol-1-phosphate (Koga,


doi: 10.1007/978-4-431-78541-5_7, 

2004). On the basis of the structure of the phospholipids as well as the structure of ribosomal RNA, many archaebacteria have been found besides the methanogens.

Some eubacteria have recently been found which have the diether-type phospholipids, the phospholipids are the derivatives of diether of *sn*-glycerol-3-phosphate (Langworthy et al., 1983). Therefore, it is still true that the phospholipids in the cell membranes of only archaea are the derivatives of diether of *sn*-glycerol-1-phosphate.

As shown in Table 7.1, many of the archaebacteria grow under extreme conditions for the organism; at very high temperatures, under very acidic or very alkaline conditions, and at very high concentrations of sodium chloride. Many of them acquire energy for the life processes from the oxidation of inorganic compounds with inorganic compounds other than oxygen or from the anaerobic decomposition of simple organic compounds such as pyruvate, though hyperhalophilic bacteria acquire the energy by the oxidation of organic compounds by oxygen.

On the basis of the 16S/18S rRNA sequence comparisons, the hyperthermophilic bacteria (optimal growth temperature higher than 80°C) are located nearer to origins of life in the phylogenetic tree (Stetter, 1994). Therefore, many researchers think that life might have originated at temperatures as high as 100°C, before the surface of the Earth had cooled to around 30°–40°C.

**Table 7.1.** Brief characteristics of several archaea [Prepared on the basis of Koga (1988) and Stetter (1994)]

| Archaea | Growth temperature (°C) | | | Growth pH | Remarks |
|---|---|---|---|---|---|
| | min. | opt. | max. | | |
| *Methanococcus jannaschii* | 50 | 85 | 86 | 3–6.5 | Acquires energy by $H_2 + CO_2 \rightarrow CH_4$. Anaerobic |
| *Methanococcus igneus* | 45 | 88 | 91 | 5–7.5 | |
| *Sulfolobus acidocaldarius* | 60 | 80 | 85 | 1–5 | Can acquire energy by the oxidation of S° with $O_2$. Aerobic |
| *Acidianus infernus* | 60 | 88 | 95 | 1.5–5 | Acquires energy by $H_2 + S° \rightarrow H_2S$. Anaerobic |
| *Pyrococcus furiosus* | 70 | 100 | 103 | 5–9 | Acquires energy by pyruvate + $H_2O \rightarrow$ acetate + $H_2 + CO_2$. Anaerobic |
| *Pyrobaculum islandicum* | 74 | 100 | 103 | 5–7 | Can acquire energy by $H_2 + S° \rightarrow H_2S$, and $H_2 + 2Fe^{3+} \rightarrow 2H^+ + 2Fe^{2+}$. Anaerobic |
| *Ferroglobus placidus* | 65 | 85 | 95 | 7 | Acquires energy by $Fe^{2+} + NO_3^- \rightarrow Fe^{3+} + NO_2^-$. Anaerobic |
| *Natronomonas pharaonis* | | 37 | | 8.5–9.5 | Acquires energy by the oxidation of organic compounds with $O_2$ in 20% NaCl. Aerobic |
| *Halobacterium salinarum* | | 30 | | 7 | Acquires energy by the oxidation of organic compounds with $O_2$ in saturated NaCl. Aerobic |

Most of the hyperthermophilic bacteria are chemolithoautotrophic (Stetter, 1994). Namely, they involve the methanogens which oxidize hydrogen with carbon dioxide, the sulfur respiratory bacteria which oxidize hydrogen with elemental sulfur, and the sulfate-reducing bacteria which oxidize hydrogen by sulfate. These results suggest that the organisms close to the origin of life may be chemolithoautotrophs. Wächthershäuser (1988) has suggested that the most primitive organisms may have acquired the energy for life processes by the reaction (7.1). However, the bacteria have not been found that utilize this reaction to acquire the energy, though one possibility can be considered that such bacteria have disappeared during the evolution of the organisms.

$$H_2S + FeS \rightarrow FeS_2 + H_2 \quad \Delta G^{\circ'} = -10\,\mathrm{kcal/mol} \tag{7.1}$$

The reaction (7.1) seems to be very important for the hydrogen utilizing bacteria to obtain hydrogen, e.g., for the methanogens and the bacteria which oxidize hydrogen by elemental sulfur. Wächthershäuser (1988) thinks that hydrogen released by the reaction (7.1) may have been used to reduce carbon dioxide to form organic compounds, but this is still very hypothetical.

## 7.2 Biological Evolution at Earlier Stages

As previously mentioned, the sulfate-reducing bacteria have been residing at least since 2750 million years ago on the basis of their growth trace detected by the isotope ratio of sulfur. Even though the bacteria are primitive, they will still have further evolved than the organisms just generated. Therefore, the origin of life will have occurred at a considerably older age than 2750 million years ago. Recent studies based on the base sequences of RNA and DNA, and other biochemical studies suggest that the life originated 3500 million or more years ago. As molecular oxygen is thought not to have been present on the primordial Earth, at least in the biosphere, the organisms of that age seem to have acquired the energy for the life processes by the anaerobic oxidation of organic and/or inorganic compounds, or by light energy.

The anaerobic decomposition of glucose to yield ATP, i.e., glycolysis or Embden–Meyerhof–Parnas pathway (Fig. 7.1a) is thought by many researchers to be the energy acquiring system which has been present since evolutionarily very old ages, because the pathway is found in a variety of the organisms, from bacteria to animals. Another decomposing pathway of glucose, the Entner–Doudoroff pathway (Fig. 7.1b), occurs in many bacteria, but this does not occur in animals. In the Embden–Meyerhof–Parnas pathway, net 2 molecules of ATP are biosynthesized per 1 molecule of glucose, while only net 1 molecule of ATP per 1 molecule of glucose is biosynthesized in the Entner–Doudoroff pathway; the efficiency of the ATP synthesis is higher in the former pathway than that in the latter. Therefore, it seems probable that the Entner–Doudoroff pathway is evolutionarily older than

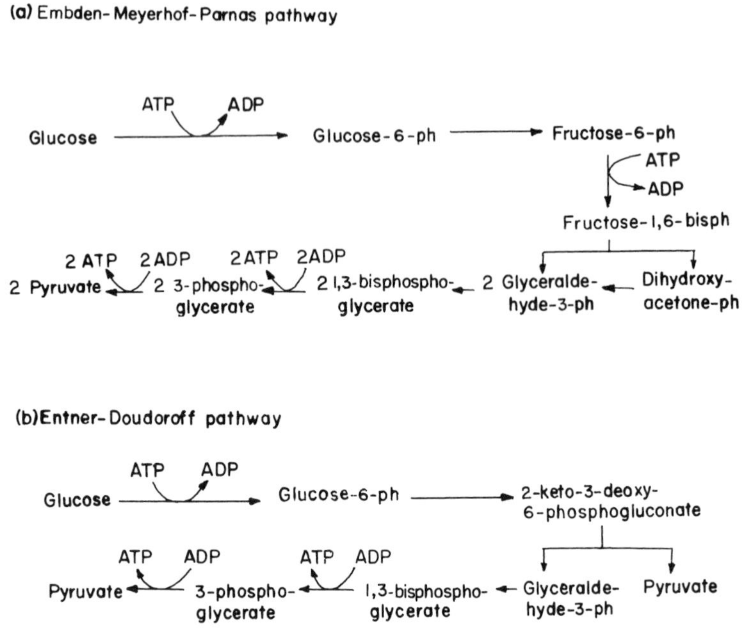


**Fig. 7.1.** Anaerobic decomposition pathways of glucose. *ph*, phosphate

the Embden–Meyerhof–Parnas pathway. At any rate, both the pathways need free glucose molecule, i.e., glucose having a free reducing group, as the starting material for the energy acquiring processes. One of the important questions is whether or not free glucose occurred on primordial Earth. It has been experimentally confirmed that amino acids can be synthesized abiotically from inorganic compounds during chemical evolution before the origin of life (Miller, 1953; and e.g. Oro, 1994). But the synthesis of free sugars together with other organic compounds, especially with amino acids, has not yet been confirmed by the experiments on chemical evolution. Although free hexoses are abiotically synthesized from formaldehyde, the reaction requires alkaline conditions. Moreover, free hexoses could not have been formed under alkaline conditions if other compounds, especially those such as amino acids, coexist. As the experiments on the synthesis of hexoses were performed using only formaldehyde, i.e., without coexistence of e.g. amino acids, free hexoses were formed.

A most important point is that free sugars (sugars having free reducing group) might have not been synthesized by the chemical evolution which required high energy such as strong irradiation of ultraviolet rays, thunder, impact by meteorites. etc., because amino acids were also formed. The amino acids will react easily with the reducing group of the free sugars when such high energies as necessary to the chemical evolution are supplied to their mixture. It is said that free sugars (especially free glucose) were biosynthesized for the first time by the photosynthesis of the green plants. These speculations mentioned above suggest that the anaerobic decomposition systems of glucose might not have occurred on the primordial Earth.

Many hyperthermophilic bacteria have been found which grow by respiration; the oxidation of hydrogen gas with carbon dioxide, with elemental sulfur, and with sulfate (Stetter, 1994). Hydrogen gas, hydrogen sulfide, and carbon dioxide derive from inside the Earth, and sulfate is thought to have occurred also on the primordial Earth as discussed below. If sulfate was present on the primordial Earth, elemental sulfur will also have been formed from hydrogen sulfide at that stage. Therefore, these bacteria which oxidize hydrogen gas with carbon dioxide, elemental sulfur, and sulfate could have resided from the earliest stage of biological evolution. Other hyperthermophilic bacteria have been found which grow by fermentation. During fermentation they anaerobically decompose pyruvate to acetate, hydrogen gas, and carbon dioxide (Stetter, 1994). As many organic compounds, such as amino acids, peptides, and lactate, seem to have occurred as the residual materials of the origin of life, the organic compounds would have been utilized as the substrate of the above fermentation. The organic compounds would have also been utilized by the sulfur respirers. For example, *Pyrobaculum islandicum* oxidizes the complex organic compounds such as yeast extract and peptone as well as hydrogen gas with elemental sulfur (Huber et al., 1987) and with ferric ion (Childers and Lovley, 2001). However, the bacterium does not utilize glucose as the starting material for the energy acquiring processes.

The results mentioned so far seem to support the idea that the energy acquiring systems using the anaerobic decomposition of glucose is not evolutionarily so old. However, a question arises as to how sulfate and ferric ion were formed under the conditions where molecular oxygen was not present. There is an idea that a photosynthesis system has evolved from a pigment system to sense infrared rays which arise from hydrothermal vents (Nisbet et al., 1995). Therefore, anoxygenic photosynthetic bacteria could have appeared around the vent at an age before the appearance of cyanobacteria. Thus, a green sulfur bacterial species has been isolated from the sea water near a hydrothermal vent at a ca. 2700 m in depth (Beatty et al., 2005). So there is a possibility that the thermophilic photosynthetic sulfur bacteria have oxidized hydrogen sulfide from the hydrothermal vent to elemental sulfur and sulfate using geothermal radiation energy in the deep sea of the primordial Earth, even if molecular oxygen was not present, though a hypothesis has been recently proposed that cyanobacteria are the origin of photosynthesis on the basis of genomic studies (Mulkidjanian et al., 2006). As the bacteria which photosynthetically oxidize ferrous ion to ferric ion under anaerobic conditions have been also found (Ehrenreich and Widdel, 1994; Heising et al., 1999), as already mentioned, ferric ion could have been formed even in the absence of molecular oxygen. Some investigators have assumed that molecular oxygen was present in the atmosphere of Earth 3500 million years ago and much sulfate was present in the ocean at that time (Lowe, 1994). The idea that sulfate, elemental sulfur, and ferric ion had been available even before the cyanobacteria appeared will be reinforced with this assumption.

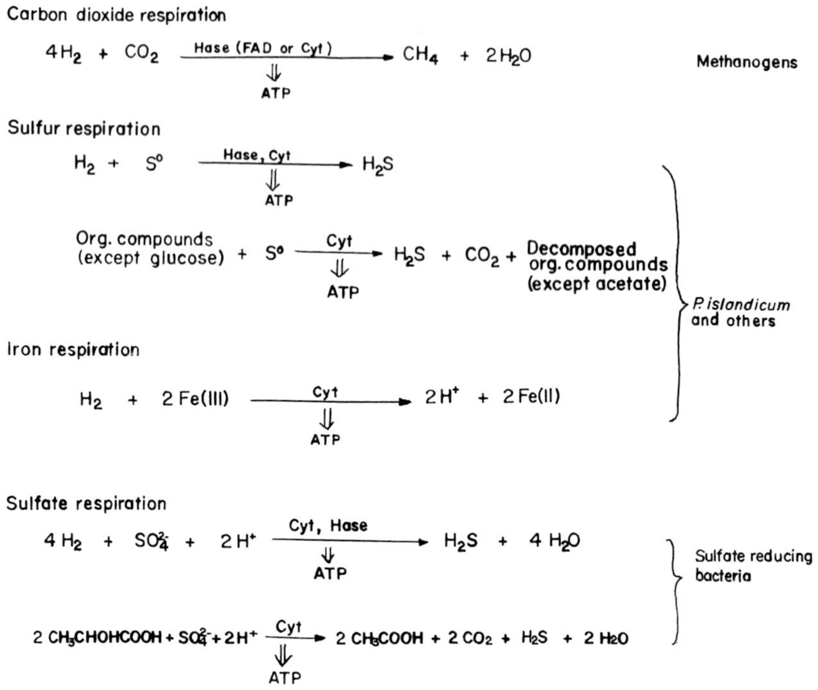


**Fig. 7.2.** The energy acquiring systems in the organisms (bacteria) which are supposed to have occurred at the earliest evolutionary stages. *Hase*, hydrogenase; *Cyt*, cytochrome. ATP is probably biosynthesized by the catalysis of ATP synthase. *ATP below downward arrow* indicates that adenosine triphosphate is biosynthesized using the energy liberated from the reactions around this indicator

The formation of methane by the methanogens is called carbon dioxide respiration, as described previously (p. 112). In general, cytochromes occur as the electron carrier in the respiration. Thus, some methanogens have cytochromes *b* and *c*. However, not all methanogens have cytochromes, as already mentioned. This may mean that carbon dioxide respiration is an early evolving stage of the respiration from the noncytochrome-type to the cytochrome-type. The electron transfer systems of the sulfur respirers contain cytochrome *b* or cytochromes *b* and *c* (Hedderich et al., 1999). In the iron respiration system of *Pyrobaculum islandicum*, cytochrome *b* seems to participate in the Fe(III) reducing system (Childers and Lovley, 2001). These results will show that the fundamental features of the respiration may have already been established in the hyperthermophilic chemolithoautotrophic archaea (Fig. 7.2).

# References


Abeliovich A, Vonshak A (1992) Anaerobic metabolism of *Nitrosomonas europaea.* Arch Microbiol 158:267–270

Adachi K, Suzuki I (1977) A study on the reaction mechanism of adenosine-5′-phosphate reductase from *Thiobacillus thioparus*, an iron-sulfur protein. Can J Biochem 55:91–98

Adams CA, Warnes GM, Nicholas DJD (1971) A sulphite-dependent nitrate reductase from *Thiobacillus denitrificans*. Biochim Biophys Acta 234:398–406

Akutsu H, Park J-S, Sano S (1993) L-Methionine methyl is specifically incorporated into the C-2 and C-7 positions of the porphyrin of cytochrome $c_3$ in a strictly anaerobic bacterium, *Desulfovibrio vulgaris.* J Am Chem Soc 115:12185–12186

Aleem MIH (1966) Generation of reducing power in chemosynthesis III. Energy-linked reduction of pyridine nucleotides in *Thiobacillus novellus.* J Bacteriol 91:729–736

Aleem MIH (1970) Oxidation of inorganic nitrogen compounds. Annu Rev Plant Physiol 21:67–90

Aleem MIH (1977) Coupling of energy with electron transfer reactions in chemolithoautotrophic bacteria. In: Haddock BA, Hamilton WA (eds) Microbial energetics. Cambridge University Press, Cambridge, pp 351–381

Aleem MIH, Hoch GE, Varner JE (1965) Water as the source of oxidant and reductant in bacterial chemosynthesis. Proc Natl Acad Sci USA 54:869–873

Aleem MIH, Lees H, Nicholas DJD (1963) Adenosine triphosphate-dependent reduction of nicotinamide adenine dinucleotide by ferro-cytochrome *c* in chemoautotrophic bacteria. Nature 200:759–761

Allen GWJ, Zinder SH (1996) Methanogenesis from acetate by cell-free extracts of the thermophilic acetotrophic methanogen *Methanothrix thermophila* (ALS-1). Arch Microbiol 166:275–281

Ambler RP (1963) The amino acid sequence of *Pseudomonas* cytochrome *c*-551. Biochem J 89:349–378

Ambler RP (1991) Sequence variability in bacterial cytochromes *c*. Biochim Biophys Acta 1058:42–47

Ambler RP, Bruschi M, KeGall J (1969) The structure of cytochrome $c_3'$ from *Desulfovibrio gigas* (NCIB 9332). FEBS Lett 5:115–117

Ambler RP, Wynn M (1973) Amino acid sequences of cytochrome *c*-551 from three species of *Pseudomonas*. Biochem J 131:485–498

ATCC (1982) American Type Culture Collection Catalogue of Strain I (15th edn.). ATCC, Rockville, MD, p 614

Aminuddin M (1980) Substrate level *versus* oxidative phosphorylation in the generation of ATP in *Thiobacillus denitrificans.* Arch Microbiol 128:19–25

Aminuddin M, Nicholas DJD (1973) Sulphide oxidation linked to the reduction of nitrate and nitrite in *Thiobacillus denitrificans*. Biochim Biophys Acta 325:81–93

Aminuddin M, Nicholas DJD (1974a) An AMP-dependent sulphite oxidase from *Thiobacillus denitrificans*: purification and properties. J Gen Microbiol 82:103–113

Aminuddin M, Nicholas DJD (1974b) Electron transfer during sulphide and sulphite oxidation in *Thiobacillus denitrificans*. J Gen Microbiol 82:115–123

Anderson IC, Poth M, Homstead J, Burdige D (1993) A comparison of NO and $N_2O$ production by autotrophic nitrifier *Nitrosomonas europaea* and the heterotrophic nitrifier *Alcaligenes faecalis*. Appl Environ Microbiol 59:3525–3533

Andersson KK, Hooper AB (1983) $O_2$ and $H_2O$ are each the source of O in $NO_2^-$ produced from $NH_3$ by *Nitrosomonas*: NMR evidence. FEBS Lett 164:236–240

Andersson KK, Lipcomb JD, Valentine M, Munck E, Hooper AB (1986) Tetraheme cytochrome *c*-554 from *Nitrosomonas europaea*. J Biol Chem 261:1126–1138

Aoki K (1978) Production method of Niter. Classical Edo Sciences Series 12 (in Japanese). Kowa-Shuppan, Tokyo

Apel WA, Dugan PR, Tuttle JH (1980) Adenosine-5′-triphosphate formation in *Thiobacillus ferrooxidans* vesicles by $H^+$ ion gradients comparable to those of environmental conditions. J Bacteriol 142:295–301

Appia-Ayme C, Bengrine A, Cavazza C, Giudici-Orticoni M, Bruschi M, Chippaux M, Bonnefoy V (1998) Characterization and expression of the co-transcribed *cyc*1 and *cyc*2 encoding the cytochrome $c_4$ (*c*-552) and a high-molecular-mass cytochrome *c* from *Thiobacillus ferrooxidans* ATCC 33020. FEBS Microbiol Lett 167:171–177

Appia-Ayme C, Guiliani N, Ratouchniak J, Bonnefoy Y (1999) Characterization of an operon encoding two *c*-type cytochromes, an $aa_3$-type cytochrome oxidase and rusticyanin in *Thiobacillus ferrooxidans* ATCC 33020. Appl Environ Microbiol 65:4781–4787

Appleby CP, Wittenberg BA, Wittenberg JB (1973) Leghemoglobin II. Changes in conformation and chemical reactivity linked to reaction with a dissociable low molecular weight ligand, X. J Biol Chem 248:3183–3187

Arciero DM, Balny C, Hooper AB (1991) Spectroscopic and rapid kinetic studies of reduction of cytochrome *c*-554 by hydroxylamine oxidoreductase from *Nitrosomonas europaea*. Biochemistry 30:11466–11472

Arciero DM, Hooper AB, Cai M, Timkovich R (1993) Evidence for structure of active site heme P-460 in hydroxylamine oxidoreductase of *Nitrosomonas*. Biochemistry 32:9370–9378

Arciero DM, Vannelli T, Logan M, Hooper AB (1989) Degradation of trichloroethylene by the ammonia-oxidizing bacterium *Nitrosomonas europaea*. Biochem Biophys Res Commun 159:640–643

Asada K (1999) The water-water cycle in chloroplasts scavenging of active oxygens and dissipation of excess photons. Annu Rev Plant Physiol Plant Mol Biol 50:601–639

Ault WU, Kulp JL (1959) Isotopic geochemistry of sulphur. Geochim Cosmochim Acta 16:201–235

Badziong W, Thauer RK (1978) Growth yields and growth rates of *Desulfovibrio vulgaris* (Marburg) growing on hydrogen plus sulfite and hydrogen plus thiosulfate as the sole energy sources. Arch Microbiol 117:209–214

Badziong W, Thauer RK, Zeikus JG (1978) Isolation and characterization of *Desulfovibrio* growing on hydrogen plus sulfate as the sole energy source. Arch Microbiol 116:41–49

Barker HA (1956) Biological formation of methane. In: Bacterial fermentation. Wiley, New York

Baross JA, Deming JW (1983) Growth of "black smoker" bacteria at temperatures of at least 250°C. Nature 303:423–426

Bashan Y, Holguin G, de-Bashan LE (2004) *Azospirillum*–plant relationships: physiological, molecular, agricultural, and environmental advances (1997–2003). Can J Microbiol 50:521–577

Bassham JA, Benson AA, Kay LD, Harris AZ, Wilson AT, Calvin M (1954) The path of carbon in photosynthesis XXI. The cyclic regeneration of carbon dioxide acceptor. J Am Chem Soc 76:1760–1770

Battersby AR (1994) New intermediates in the $B_{12}$ pathway. In: The biosynthesis of the tetrapyrrole pigments (Ciba Foundation Symposium 180) Wiley, Chichester, pp 267–279

Baulcombe D, Verma DP (1978) Preparation of complementary DNA for leghemoblobin and direct demonstration that leghemoglobin encoded by the soybeangenome. Nucleic Acids Res 5:4141–4155

Baumgarten A, Redenius I, Kranczoch J, Cypionka H (2001) Periplasmic oxygen reduction by *Desulfovibrio* species. Arch Microbiol 176:306–309

Beatty JT, Overmann J, Lince MT, Manske AK, Lang AS, Blankenship RE, Van Dover CL, Martinson TA, Plumley FG (2005) An obligately photosynthetic bacterial anaerobe from a deep-sea hydrothermal vent. Proc Natl Acad Sci USA 102:9306–9310

Beaumont HJ, Hommes NG, Sayavedra-Soto LA, Arp DJ, Arciero DM, Hooper AB, Westerhoff HV, van Spanning RJ (2002) Nitrite reductase of *Nitrosomonas europaea* is not essential for production of gaseous nitrogen oxides and confers tolerance to nitrite. J Bacteriol 184:2557–2560

Bédard C, Knowles R (1989) Physiology, biochemistry, and specific inhibitors of $CH_4$, $NH_4^+$, and CO oxidation by methanotrophs and nitrifiers. Microbiol Rev 53:68–84

Beffa T, Fischer C, Aragno M (1993) Growth and respiratory oxidation of reduced sulfur compounds by intact cells of *Thiobacillus novellus* (type strain) grown on thiosulfate. Curr Microbiol 26:323–326

Beller HR, Chain PSG, Letain TE, Chakicherla A, Larimer FW, Richardson PM, Coleman MA, Wood AP, Kelly DP (2006) The genome sequence of the obligately chemolithoautotrophic, facultatively anaerobic bacterium *Thiobacillus denitrificans*. J Bacteriol 188:1473–1488

Bengrine A, Guiliani N, Appia-Ayme C, Jedlicki E, Holmes DS, Chippaux M, Bonnefoy V (1998) Sequence and expression of the rusticyanin structural gene from *Thiobacillus ferrooxidans* ATCC 33020 strain. Biochim Biophys Acta 1443:99–112

Benosman H, Asso M, Bertrand P, Yagi T, Gayda J-P (1989) EPR study of the redox interactions in cytochrome $c_3$ from *Desulfovibrio vulgaris* Miyazaki. Eur J Biochem 82:51–55

Berben G (1996) *Nitrobacter winogradskyi* cytochrome *c* oxidase gene are organized in a repeated gene cluster. Antonie von Leeuwenhoek 69:305–315

Bergmann DJ, Arciero DM, Hooper AB (1994) Organization of the *hao* gene cluster of *Nitrosomonas europaea*: genes for two tetraheme C cytochromes. J Bacteriol 176:3148–3153

Bergmann DJ, Hooper AB (1994a) Sequence of the gene, *amo*B, for the 43 kDa polypeptide of ammonia monooxygenase of *Nitrosomonas europaea.* Biochem Biophys Res Commun 204:759–762

Bergmann DJ, Hooper AB (1994b) The primary structure of cytochrome P-460 of *Nitrosomonas europaea*: presence of a c-heme binding motif. FEBS Lett 353:324–326

Berk H, Thauer RK (1997) Function of $F_{420}$-dependent NADP reductase in methanogenic archaea containing an NADP-dependent alcohol dehydrogenase. Arch Microbiol 168:396–402

Bertsova YV, Bogachev AV, Skulachev VP (2001) Noncoupled NADH: ubiquinone oxidoreductase of *Azotobacter vinelandii* is required for diazotrophic growth at high oxygen concentration. J Bacteriol 183:6869–6874

Bhatnagar L, Jain MK, Zeikus JG (1991) Methanogenic bacteria. In: Shively JM, Barton LL (eds) Variations in autotrophic life. Academic Press, London, pp 251–270

Blake RC Jr, Howard GT, McGinness S (1994) Enhanced yields of iron-oxidizing bacteria by in situ electrochemical reduction of soluble iron in the growth medium. Appl Environ Microbiol 60:2704–2710

Blake RC Jr, McGinness S (1993) Electron-transfer proteins of bacteria that respire on iron. In: Torma AE, Apel, ML, Brierly CL (eds) Biohydrometallurgical technologies, vol. II. Minerals, Metals, Materials Society, Warrendale, PA, pp 615–628

Blake RC Jr, Shute EA (1987) Respiratory enzymes of *Thiobacillus ferrooxidans.* A kinetic study of electron transfer between iron and rusticyanin in sulfate media. J Biol Chem 262: 14983–14989

Blake RC Jr, Shute EA (1994) Respiratory enzymes of *Thiobacillus ferrooxidans.* Kinetic properties of an acid-stable iron rusticyanin oxidoreductase. Biochemistry 33:9220–9228

Blake RC Jr, Shute EA, Greenwood MM, Spencer GH, Ingledew WJ (1993) Enzymes of aerobic respiration on iron. FEMS Microbiol Rev 11:9–18

Blakemore RP (1982) Magnetotactic bacteria. Annu Rev Microbiol 36:217–238

Blaut M, Gottschalk G (1984) Coupling of ATP synthesis and methane formation from methanol and molecular hydrogen in *Methanosarcina barkeri*. Eur J Biochem 141:217–222

Blaut M, Gottschalk G (1985) Evidence for a chemiosmotic mechanism of ATP synthesis in methanogenic bacteria. Trends Biochem Sci 10:486–489

Bock E, Schmidt I, Stuven R, Zart D (1995) Nitrogen loss caused by denitrifying *Nitrosomonas* cells using ammonium or hydrogen as electron donors and nitrate as electron acceptor. Arch Microbiol 163:16–20

Bodo C, Lundgren DG (1974) Iron oxidation by cell envelopes of *Thiobacillus ferrooxidans*. Can J Microbiol 20:1647–1652

Bosecker K (1997) Bioleaching: metal solubilization by microorganisms. FEMS Microbiol Rev 20:591–604

Botuyan MV, Toy-Palmer A, Chung J, Blake RC Jr, Beroza P, Case DA, Dyson HJ (1996) NMR solution structure of Cu(I) rusticyanin from *Thiobacillus ferrooxidans*: structural basis for the extreme acid stability and redox potential. J Mol Biol 263:752–767

Bramblett RN, Peck HD Jr (1975) Some physical and kinetic properties of adenylyl sulfate reductase from *Desulfovibrio vulgaris*. J Biol Chem 250:2979–2986

Brasseur G, Bruscella P, Bonnefoy V, Lamesle-Meunier D (2002) The $bc_1$ complex of the iron-grown acidophilic chemolithotrophic bacterium *Acidithiobacillus ferrooxidans* in the reverse but not in the forward direction. Is there a second $bc_1$ complex? Biochim Biophys Acta 1555:37–43

Brierley JA, Lockwood SJ (1977) The occurrence of thermophilic iron-oxidizing bacteria in a copper leaching system. FEMS Microbiol Lett 2:163–165

Brill WJ (1980) Biochemical genetics of nitrogen fixation. Microbiol Rev 44:449–467

Brock TD, Schlegel HG (1989) Introduction. In: Schlegel HG, Bowien B (eds) Autotrophic bacteria. Science Tech Publishers, Madison, WI/Springer, Berlin, pp 1–15

Brodersen J, Gottschalk G, Deppenmeier U (1999) Membrane-bound $F_{420}H_2$-dependent heterosulfide reduction in *Methanococcus voltae*. Arch Microbiol 171:115–121

Brown LD, Dennehy ME, Rawlings DE (1994) The $F_1$ genes of $F_1F_o$ATP synthase from the acidophilic bacterium *Thiobacillus ferrooxidans* complement *Escherichia coli* $F_1$ *unc* mutants. FEMS Microbiol Lett 122:19–26

Bruscella P, Cassagnaud L, Ratouchniak J, Brasseur G, Lojou E, Amils R, Bonnefoy V (2005) The HiPIP from the acidophilic *Acidithiobacillus ferrooxidans* is correctly processed and translocated in *Escherichia coli*, in spite of the periplasm pH difference between these two micro-organisms. Microbiology 151:1421–1431

Bruschi M, Leroy G, Bonicel J, Campese D, Dolla A (1996) The cytochrome $c_3$ family: amino acid sequence of dimeric octahaem cytochrome $c_3$ ($M_r$ 26000) isolated from *Desulfovibrio gigas*. Biochem J 320:933–938

Bryant MP, Wolin EA, Wolin MJ, Wolfe RS (1967) *Methanobacillus omelianskii*, a symbiotic association of two species of bacteria. Arch Microbiol 59:20–31

Cammack R, Chapman A, Lu W-P, Kargouni A, Kelly DP (1989) Evidence that protein B of the thiosulfate-oxidizing system of *Thiobacillus versutus* contains a binuclear manganese cluster. FEBS Lett 253:239–243

Cantera JJL, Stein LY (2007) Role of nitrite reductase in the ammonia-oxidizing pathway of *Nitrosomonas europaea*. Arch Microbiol 188:349–354

Carr GJ, Ferguson SJ (1990) The nitric oxide reductase of *Paracoccus denitrificans*. Biochem J 269:432–429

Castelfranco PA, Walker CJ, Weinstein JD (1994) Biosynthetic studies on chlorophylls: from protoporphyrin IX to protochlorophyllide. In: The biosynthesis of the tetrapyrrole pigments (Ciba Foundation Symposium 180). Wiley, Chichester, pp 194–204

Castignetti D, Hollocher TC (1982) Nitrogen redox metabolism of a heterotrophic, nitrifying-denitrifying *Alcaligenes* sp. from soil. Appl Environ Microbiol 44:923–928

Castignetti D, Paulutsis D, Turley J (1990) An examination of protein translocation and energy conservation during heterotrophic nitrification. FEMS Microbiol Lett 66:175–182

Castignetti D, Petithory JR, Hollocher TC (1983) Pathway of oxidation of pyruvic oxime by a heterotrophic nitrifier of the genus *Alcaligenes*: evidence against hydrolysis to pyruvate and hydroxylamine. Arch Biochem Biophys 224:587–593

Caughey WS (1971) Structure-function relationships in cytochrome *c* oxidase and other hemoproteins. Adv Chem Ser 100:248–269

Cavalcante VA, Dobereiner J (1988) A new-tolerant nitrogen-fixing bacterium associated with sugarcane. Plant Soil 108:23–31

Cavazza C, Bruschi M (1995) Iron oxidation by *Thiobacillus ferrooxidans*: characterization of two electron transfer proteins. In: Jerez CA, Varges T, Toledo H, Wiertz JV (eds) Biohydrometallurgical processing, vol II. University of Chile, Santiago, pp 97–107

Cavazza C, Giudici-Orticoni M-T, Nitschle W, Appia C, Bonnefoy V, Bruschi M (1996) Characterization of a soluble cytochrome $c_4$ isolated from *Thiobacillus ferrooxidans*. Eur J Biochem 242:308–314

Cerdá J, González S, Ríos JM, Quintana T (1993) Uranium concentrates bioproduction in Spain: a case study. FEMS Microbiol Rev 11:253–260

Chan CW, Suzuki I (1994) Thiosulfate oxidation by sulfur-grown *Thiobacillus thiooxidans* cells, cell-free extracts, and thiosulfate-oxidizing enzyme. Can J Microbiol 40:816–822

Chan L, Pereira MM, Xavier AV, LeGall J (1994) Isolation and characterization of a high molecular weight cytochrome from the sulfate reducing bacterium *Desulfovibrio gigas*. FEBS Lett 347:295–299

Chan MK, Kim J, Rees DC (1993) The nitrogenase FeMo-cofactor and P-cluster: 22Å resolution structures. Science 260:792–794

Chang CK, Timkovich R, Wu W (1986) Evidence that heme $d_1$ is a 1,3-porphyrindione. Biochemistry 25:8477–8453

Charles AM, Suzuki I (1966a) Mechanism of thiosulfate oxidation by *Thiobacillus novellus*. Biochim Biophys Acta 128:510–521

Charles AM, Suzuki I (1966b) Purification and properties of sulfite:cytochrome *c* oxidoreductase from *Thiobacillus novellus*. Biochim Biophys Acta 128:522–534

Charnock JM, Dreusch A, Körner H, Neese F, Nelson J, Kannt A, Michel H, Garner CD, Kroneck PM, Zumft WG (2000) Structural investigations of the CuA centre of nitrous oxide reductase from *Pseudomonas stutzer* by site-directed mutagenesis and X-ray absorption spectroscopy. Eur J Biochem 267:1366–1381

Chaudhry GR, MacGregor CH (1983) Cytochrome *b* from *Escherichia coli* nitrate reductase. Its properties and association with the enzyme complex. J Biol Chem 258:5819–5827

Chaudhry GR, Suzuki I, Lees H (1980) Cytochrome oxidase of *Nitrobacter agilis*: isolation by hydrophobic chromatography. Can J Microbiol 26:1270–1274

Chaudhuri SK, Lack JG, Coates JD (2001) Biogenic magnetite formation through anaerobic biooxidation of Fe(II). Appl Environ Microbiol 67:2844–2848

Cheeseman P, Toms-Wood A, Wolfe RS (1972) Isolation and properties of a fluorescent-compound, Factor 420, from *Methanobacterium* strain M.o.H. J Bacteriol 112:527–531

Childers SE, Lovley DR (2001) Differences in Fe(III) reduction in the hyperthermophilic archaeon, *Pyrobaculum islandicum*, versus mesophilic Fe(III)-reducing bacteria. FEMS Microbiol Lett 195:253–258

Childress JJ, Felbeck H, Somero GN (1987) Symbiosis in the deep sea. Sci Am 256:106–112

Claassen R, Logan CT, Snyman CP (1993) Bio-oxidation of refractory goldbearing arsenopyritic ores. In: Torma AE, Wey JE, Lakshmann VI (eds) Biohydrometallurgical technologies, vol I. Minerals, Metals & Materials Society, Warrendale, PA, pp 479–488

Coates JD, Bhupathireja VK, Achenbach LA, McInerney MJ, Lovley DR (2001) *Geobacter hydrogenophilus*, *Geobacter chapelli* and *Geobacter grbiciae*, three new, strictly anaerobic, dissimilatory Fe(III)-reducers. Int J Syst Evol Microbiol 51:581–588

Cobley JG (1976) Energy-conserving reactions in phosphorylating electron-transport particles from *Nitrobacter winogradskyi*. Activation of nitrite oxidation by electrical component of the proton motive force. Biochem J 156:481–491

Cobley JG (1984) Oxidation of nitrite and formate in *Nitrobacter* membrane preparations: evidence that both reactions are catalyzed by the same enzyme. In: Strohl WR, Tuovinen OH (eds) Microbial chemoautotrophy. Ohio State University Press, Columbus, pp 169–183

Cobley JG, Haddock BA (1975) The respiratory chain of *Thiobacillus ferrooxidans*: reduction of cytochrome by $Fe^{2+}$ and preliminary characterization of rusticyanin, a novel "blue" copper protein. FEBS Lett 60:29–33

Collinet M-N, Morin D (1990) Characterization of arsenopyrite oxidizing *Thiobacillus*. Tolerance to arsenite, arsenate and ferric iron. Antonie van Leeuwenhoek 57:237–244

Collins MJ, Arciero DM, Hooper AB (1993) Optical spectropotentiometric resolution of the hemes of hydroxylamine oxidoreductase. Heme quantitation and pH dependence of $E_m$. J Biol Chem 268:14655–14662

Corbett CM, Ingledew WJ (1987) Is $Fe^{3+}/Fe^{2+}$ cycling an intermediate in sulphur oxidation by *Thiobacillus ferrooxidans*? FEMS Microbiol Lett 41:1–6

Cox JC, Boxer DH (1978) The purification and some properties of rusticyanin, a blue copper protein involved in iron (II) oxidation from *Thiobacillus ferro-oxidans*. Biochem J 174:497–502

Coyne MS, Arunakumari A, Averill BA, Tiedje JM (1989) Immunological identification and distribution of dissimilatory heme $cd_1$ and nonheme copper nitrite reductase in denitrifying bacteria. Appl Environ Microbiol 55:2924–2931

Cutting JA, Schulman HM (1969) The site of heme synthesis in soybean root nodules. Biochim Biophys Acta 192:486–493

Czjzek M, Payan F, Guerlesquin F, Bruschi M, Haser R (1994) Crystal structure of cytochrome $c_3$ from *Desulfovibrio desulfuricans* Norway at 1.7Å resolution. J Mol Biol 243:653–667

Daniels L, Sparling R, Sprott D (1984) The bioenergetics of methanogenesis. Biochim Biophys Acta 768:113–163

Das A, Mishra AK, Roy P (1992) Anaerobic growth on elemental sulfur using dissimilar iron reduction by autotrophic *Thiobacillus ferrooxidans*. FEMS Microbiol Lett 9:167–172

de Bruijn P, van de Graaf AA, Jetten MSM, Robertson LA (1995) Growth of *Nitrosomonas europaea* on hydroxylamine. FEMS Microbiol Lett 125:179–184

de Jong GAH, Hazeu V, Bos P, Kuenen JG (1997) Isolation of the tetrathionate hydrolase from *Thiobacillus acidophilus*. Eur J Biochem 243:678–683

de Vries S, Strampraad MJ, Lu S, Moënne-Loccoz P, Schröder I (2003) Purification and characterization of $QH_2$:NO reductase from hyperthermophilic archaeon *Pyrobaculum aerophilum*. J Biol Chem 278:35861–35868

de Wit R, van Gemerden H (1987) Oxidation of sulfide to thiosulfate by *Microcoleus chthonoplastes*. FEMS Microbiol Ecol 45:7–13

Dickerson RE, Timkovich RC (1975) Cytochromes *c*. In: Boyer PD (ed) The enzymes, vol XI part A. Academic Press, New York, pp 397–547

Diekert G, Klee B, Thauer RK (1980) Nickel, a component of factor $F_{430}$ from *Methanobacterium thermoautotrophicum*. Arch Microbiol 124:103–106

DiSpirito AA, Lipscomb JD, Hooper AB (1986) Cytochrome $aa_3$ from *Nitrosomonas europaea*. J Biol Chem 261:17048–17056

DiSpirito AA, Taaffe LR, Lipscomb JD, Hooper AB (1985) A "blue" copper oxidase from *Nitrosomonas europaea*. Biochim Biophys Acta 827:320–326

Djebli A, Proctor P, Blake RC Jr, Shoham M (1992) Crystallization and preliminary X-ray crystallographic studies of rusticyanin from *Thiobacillus ferrooxidans*. J Mol Biol 227:581–582

Dua RD, Bhandan B, Nicholas DJD (1979) Stable isotope studies on the oxidation of ammonia to hydroxylamine by *Nitrosomonas europaea*. FEBS Lett 106:401–404

Dymond J, Collier RW, Watwood ME (1989) Bacterial mats from Crater Lake, Oregon and their relationship to possible deep-lake hydrothermal venting. Nature 342:673–675

Eady RR (1988) The vanadium-containing nitrogenase of *Azotobacter*. Biofactors 1:111–116

Eccleston M, Kelly DP, Wood AP (1985) Autotrophic growth and iron oxidation and inhibition kinetics of *Leptospirillum ferrooxidans*. In: Cadwell DE, Brierley JA, Brierley CL (eds) Planetary ecology. van Nostrand Reinhold, New York, pp 263–272

Ehrenreich A, Widdel F (1994) Anaerobic oxidation of ferrous iron by purple bacteria, a new type of phototrophic metabolism. Appl Environ Microbiol 60:4517–4526

Ehrlich HL (1984) Different forms of bacterial manganese oxidation. In: Strohl WR, Tuovinen OH (eds) Microbial chemoautotrophy. Ohio State University Press, Columbus, pp 47–56

Ehrlich HL, Ingledew WJ, Salerno JC (1991) Iron- and manganese-oxidizing bacteria. In: Shively JM, Barton LL (eds) Variations in autotrophic life. Academic Press, London, pp 147–170

Eirich LD, Vogels GD, Wolfe RS (1978) Proposed structure for coenzyme $F_{420}$ from *Methanobacterium.* Biochemistry 17:4583–4593

Elbehti A, Brasseur G, Lemesle-Meunier D (2000) First evidence for existence of an uphill electron transfer through the $bc_1$ and NADH-Q oxidoreductase complexes of the acidophilic obligate chemolithotrophic ferrous iron-oxidizing bacterium *Thiobacillus ferrooxidans.* J Bacteriol 182:3602–3606

Elbehti A, Lemesle-Meunier D (1996) Identification of membrane-bound *c*-type cytochromes in an acidophilic ferrous ion oxidizing bacterium *Thiobacillus ferrooxidans*. FEMS Microbiol Lett 136:51–56

Ellermann J, Hedderich R, Bocher R, Thauer RK (1988) The final step in methane formation. Investigations with highly purified methyl-CoM reductase (component C) from *Methanobacterium thermoautotrophicum* (strain Marburg). Eur J Biochem 172:669–677

Ellfolk N, Sievers G (1974) Correction of the amino acid sequence of soybean leghemoglobin alpha. Acta Chem Scand B 28:1245–1246

Elstner EF, Stoffer C, Heupel A (1975) Determination of superoxide free radical ion and hydrogen peroxide as products of photosynthetic oxygen reduction. Z Naturforsch 30c:53–57

Emmel T, Sand W, Konig WA, Bock E (1986) Evidence for the existence of a sulphur oxygenase in *Sulfolobus brierleyi.* J Gen Microbiol 132:3415–3420

Ensign SA, Hyman MR, Arp DJ (1993) *In vitro* activation of ammonia monooxygenase from *Nitrosomonas europaea* by copper. J Bacteriol 175:1971–1980

Erickson RH, Hooper AB (1972) Preliminary characterization of variant CO-binding heme protein from *Nitrosomonas.* Biochim Biophys Acta 275:231–244

Erickson RH, Hooper AB, Terry KR (1972) Solubilization and purification of cytochrome $a_1$ from *Nitrosomonas.* Biochim Biophys Acta 283:155–166

Evans MCW, Buchanan BB, Arnon DI (1966) A new ferredoxin-dependent carbon reduction cycle in a photosynthetic bacterium. Proc Natl Acad Sci USA 55:928–934

Falk JE (1964) Porpyrins and metalloporphyrins. Elsevier, Amsterdam

Fareleira P, LeGall J, Xavier AV, Santos H (1997) Pathways for utilization of carbon reserves in *Desulfovibrio gigas* under fermentative and respiratory conditions. J Bacteriol 179:3972–3980

Fauque G, Herve D, LeGall J (1979) Structure–function relationship in hemoproteins: the role of cytochrome $c_3$ in the reduction of colloidal sulfur by sulfate-reducing bacteria. Arch Microbiol 121:261–264

Fauque G, LeGall J, Barton LL (1991) Sulfate-reducing and sulfur-reducing bacteria. In: Shively JM, Barton LL (eds) Variations in autotrophic life. Academic Press, London, pp 271–337

Fay P (1992) Oxygen relations of nitrogen fixation in cyanobacteria. Microbiol Rev 56:340–373

Fay P, Stewart WDP, Walsby AE, Fogg GE (1968) Is the heterocyst the site of nitrogen fixation in blue-green algae? Nature 220:810–812

Ferry JG (1992) Methane from acetate. J Bacteriol 174:5489–5495

Fischer R, Thauer RK (1988) Methane formation from acetyl phosphate in cell extracts of *Methanosarcina barkeri.* Dependence of the reaction on coenzyme A. FEBS Lett 228:249–253

Fisher JR (1966) Bacterial leaching of Elliot Lake uranium ore. Transactions 69:167–171

Freitag A, Bock E (1990) Energy conservation in *Nitrobacter.* FEMS Microbiol Lett 66:157–162

Froud SJ, Anthony C (1984) Purification and characterization of the *o*-type cytochrome oxidase from *Methylophilus methylotrophus*, and its reconstitution into methanol oxidase electron transport chain. J Gen Microbiol 130:2201–2212

Fry IN, Lazaroff N, Packer L (1986) Sulfate-dependent iron oxidation by *Thiobacillus ferrooxidans*: characterization of new EPR detectable electron transport component on the reducing side of rusticyanin. Arch Biochem Biophys 246:650–654

Fuchs G, Stupperich E (1978) Evidence for an incomplete reductive carboxylic acid cycle in *Methanobacterium thermoautotrophicum*. Arch Microbiol 118:121–125

Fuchs G, Stupperich E, Eden G (1980a) Autotrophic $CO_2$ fixation in *Chlorobium limicola*. Evidence for operation of a reductive tricarboxylic acid cycle in growing cells. Arch Microbiol 128:64–71

Fuchs G, Stupperich E, Jaenchen R (1980b) Autotrophic $CO_2$ fixation in *Chlorobium limicola*. Evidence against the operation of the Calvin cycle in growing cells. Arch Microbiol 128:56–63

Fujiwara T, Fukumori Y (1996) Cytochrome *cb*-type nitric oxide reductase with cytochrome *c* oxidase activity from *Paracoccus denitrificans* ATCC 35512. J Bacteriol 176:1866–1871

Fujiwara T, Fukumori Y, Yamanaka T (1992) A novel terminal oxidase, cytochrome $baa_3$ purified from aerobically grown *Pseudomonas aeruginosa*: it shows clear difference between resting state and pulsed state. J Biochem 112:290–298

Fujiwara T, Yamanaka T, Fukumori Y (1995) The amino acid sequence of *Nitrosomonas europaea* cytochrome *c*-552. Curr Microbiol 31:1–4

Fukaya M, Tayama K, Tamaki T, Ebisuya H, Okumura H, Kawamura Y, Horinouchi S, Beppu T (1993) Characterization of a cytochrome $a_1$ that functions as a ubiquinol oxidase in *Acetobacter aceti*. J Bacteriol 175:4307–4314

Fukumori Y (2000) Structure and function of nano-size biomagnetic particles. Seikagaku 72:1165–1168

Fukumori Y, Hoshiko K, Yamanaka T (1989) Purification and some properties of thiosulfate-cleaving enzyme from *Thiobacillus novellus*. FEMS Microbiol Lett 65:159–164

Fukumori Y, Saito T, Yamazaki T, Yamanaka T (1988a) Purification and some properties of cytochrome *b*-560 from *Nitrosomonas europaea*. Plant Cell Physiol 29:1077–1079

Fukumori Y, Yamanaka T (1979) Flavocytochrome *c* of *Chromatium vinosum*: some enzymatic properties and subunit structure. J Biochem 85:1405–1414

Fukumori Y, Yano T, Sato A, Yamanaka T (1988b) Fe(II)-oxidizing enzyme purified from *Thiobacillus ferrooxidans*. FEMS Microbiol Lett 50:169–172

Fukuoka M, Fukumori Y, Yamanaka T (1987) *Nitrobacter winogradskyi* cytochrome $a_1c_1$ is an iron-sulfur molybdoenzyme having hemes *a* and *c*. J Biochem 102:525–530

Fülöp V, Moir JW, Ferguson S, Hajdu J (1995) The anatomy of a bifunctional enzyme structural basis for reduction of oxygen to water and synthesis of nitric oxide by cytochrome $cd_1$. Cell 81:369–377

Furukawa Y, Inubushi K (2002) Feasible suppression technique of methane emission from paddy soil by iron amendment. Nutr Cycl Agroecosyst 64:193–201

Fuseler K, Krekeler D, Sydow U, Cypionka H (1996) A common pathway of sulfide oxidation by sulfate-reducing bacteria. FEMS Microbiol Lett 144:129–134

Galushko AS, Schink B (2000) Oxidation of acetate through reactions of the citric acid cycle by *Geobacter sulfurreducens* in pure culture and in syntrophic coculture. Arch Microbiol 174:314–321

Garcia-Horsman JA, Berry E, Shapleigh JP, Alben JO, Gennis RB (1994) A novel cytochrome *c* oxidase from *Rhodobacter sphaeroides* that lacks CuA. Biochemistry 33:3113–3119

Georgiadis MM, Komiya H, Chakrabarti P, Woo D, Kornuc JJ, Rees DC (1992) Crystallographic structure of nitrogenase iron protein from *Azotobacter vinelandii*. Science 257:1653–1659

Giudici-Orticoni M-T, Leroy G, Nitschke W, Bruschi M (2000) Characterization of a new dihemic $c_4$-type cytochrome isolated from *Thiobacillus ferrooxidans*. Biochemistry 39:7205–7211

Grahame DA, Stadtman TC (1987) *In vitro* methane and methyl coenzyme M formation from acetate: evidence that acetyl-CoM is the required intermediate activated form of acetate. Biochem Biophys Res Commun 147:254–258

Grassle JF (1985) Hydrothermal vent animals: distribution and biology. Science 229:713–717

Gray KA, Grooms M, Myllykallio H, Moomaw C, Slaughter C, Daldal F (1994) *Rhodobacter capsulatus* contains a novel *cb*-type cytochrome *c* oxidase without a CuA center. Biochemistry 33:3120–3127

Guerlesquin F, Bovier-Lapierre G, Bruschi M (1982) Purification and characterization of cytochrome $c_3$ (*Mr* 26000) isolated from *Desulfovibrio desulfuricans* Norway strain. Biochem Biophys Res Commun 105:530–538

Gunsalus RP, Wolfe RS (1978) Chromophoric factors $F_{432}$ and $F_{430}$ of *Methanobacterium thermoautotrophicum*. FEMS Microbiol Lett 3:191–193

Hafenbradl D, Keller M, Dirmeier R, Rachel R, Rossnagel P, Burggraf S, Huber H, Stetter KO (1996) *Ferroglobus placidus* gen. nov., a novel hyperthermophilic archaeum that oxidizes $Fe^{2+}$ at neutral pH under anoxic conditions. Arch Microbiol 166:308–314

Hall GH (1986) Nitrification in lakes. In: Prosser JI (ed) Nitrification. IRL Press, Oxford, pp 127–156

Hallbeck L, Pedersen K (1991) Autotrophic and mixotrophic growth of *Gallionella ferruginea*. J Gen Microbiol 137:2657–2661

Hallberg KB, Dopson M, Lindström EB (1996) Reduced sulfur compound oxidation by *Thiobacillus caldus*. J Bacteriol 178:6–11

Hanert HH (1989) Nonbudding, stalked bacteria. Genus *Gallionella feruginea* Ehrenberg 1838. In: Stanley JT, Bryant MP, Pfennig N, Holt JG (eds) Bergey's manual of systematic bacteriology vol 3. Williams & Wilkins, Baltimore, pp 1974–1979

Hara T, Vallalobos AP, Fukumori Y, Yamanaka T (1991) Purification and characterization of ATPase from *Nitrobacter winogradskyi*. FEMS Microbiol Lett 82:49–54

Harahuc L, Lizama HM, Suzuki I (2000) Selective inhibition of the oxidation of ferrous iron or sulfur in *Thiobacillus ferrooxidans*. Appl Environ Microbiol 66:1031–1037

Harahuc L, Suzuki I (2001) Sulfite oxidation by iron-grown cells of *Thiobacillus ferrooxidans* at pH 3 possibly involves free radicals, iron, and cytochrome oxidase. Can J Microbiol 47:424–430

Harder SR, Lu W-P, Freiberg BA, Ragsdale SW (1989) Spectroelectrochemical studies of corrinoid iron-sulfur protein involved in acetyl coenzyme-A synthesis by *Clostridium thermoaceticum*. Biochemistry 28:9080–9087

Hart A, Murrel JC, Poole RK, Norris PR (1991) An acid-stable cytochrome in iron-oxidizing *Leptospirillum ferrooxidans*. FEMS Microbiol Lett 81:89–94

Hatch MD, Slack CR, Johnson HS (1967) Further studies on a new pathway of photosynthetic carbon dioxide fixation in sugar-cane and its occurrence in other plant species. Biochem J 102:417–422

Hatchikian EC, LeGall J, Bruschi M, Dubourdieu M (1972) Regulation of the reduction of sulfite and thiosulfate by ferredoxin, flavodoxin and cytochrome $cc_3$ in extracts of the sulfate reducer *Desulfovibrio gigas*. Biochim Biophys Acta 258:701–708

Hazeu W, Schmedding DJM, Goddijn O, Bos P, Kuenen JG (1987) The importance of the sulphur-oxidizing capacity of *Thiobacillus ferrooxidans* during leaching of pyrite. Proceedings of the 4th European Congress in Biotechnology vol. 3. Amsterdam, pp 497–499

Hazra TK, Mukherjea M, Mukherjea RN (1992) Role of rusticyanin in the electron transport process in *Thiobacillus ferrooxidans*. Indian J Biochem Biophys 29:77–81

Hedderich R, Kimmek O, Kröger A, Dirmeier R, Keller M, Stetter KO (1999) Anaerobic respiration with elemental sulfur and with disulfides. FEMS Microbiol Rev 22:353–381

Heiden S, Hedderich R, Setzke E, Thauer RK (1993) Purification of a cytochrome *b* containing $H_2$:heterosulfide oxidoreductase of *Methanosarcina barkeri*. Eur J Biochem 213:529–535

Heiden S, Hedderich R, Setzke E, Thauer RK (1994) Purification of a two-subunit cytochrome *b*-containing heterosulfide reductase from methanol-grown *Methanosarcina barkeri*. Eur J Biochem 221:855–861

Heising S, Richter L, Ludwig W, Schink B (1999) *Chlorobium ferrooxidans* sp. nov., a phototrophic green sulfur bacterium that oxidizes ferrous iron in coculture with a "*Geospirillum*" sp. strain. Arch Microbiol 172:116–124

Heiss B, Frunzke K, Zumft WG (1989) Formation of the N-N bond from nitric oxide by a membrane-bound cytochrome *bc* complex of nitrate-respiring (denitrifying) *Pseudomonas stutzeri*. J Bacteriol 171:3288–3297

Higuchi Y, Bando S, Kusunoki M, Matsuura Y, Yasuoka N, Kakudo M, Yamanaka T, Yagi T, Inokuchi H (1981) The structure of cytochrome $c_3$ from *Desulfovibrio vulgaris* Miyazaki at 2.5 Å resolution. J Biochem 89:1659–1662

Higuchi Y, Kusunoki M, Matsuura Y, Yasuoka N, Kakudo M (1984) Refined structure of cytochrome $c_3$ at 1.8 Å resolution. J Mol Biol 172:109–139

Higuchi Y, Inaka K, Yasuoka N, Yagi T (1987) Isolation and crystallization of high molecular weight cytochrome from *Desulfovibrio vulgaris* Hildenborough. Biochim Biophys Acta 911:341–348

Hofman T, Lees H (1953) The biochemistry of nitrifying organisms 4. The respiration and intermediary metabolism of *Nitrosomonas*. Biochem J 54:579–583

Holliger C, Wohlfarth G, Diekert G (1999) Reductive dehalogenation in the energy metabolism of anaerobic bacteria. FEMS Microbiol Rev 22:383–398

Hollocher TC, Tate ME, Nicholas DJD (1981) Oxidation of ammonia by *Nitrosomonas europaea*: definitive $^{18}$O-tracer evidence that hydroxylamine formation involves a monooxygenase. J Biol Chem 256:10834–10836

Hollocher TC, Kumar S, Nicholas DJD (1982) Respiration-dependent proton translocation in *Nitrosomonas* and its apparent absence in *Nitrobacter agilis* during inorganic oxidation. J Bacteriol 149:1013–1020

Hooper AB (1968) A nitrite reducing enzyme from *Nitrosomonas europaea*. Preliminary characterization with hydroxylamine as electron donor. Biochim Biophys Acta 162:49–65

Hooper AB, Erickson RH, Terry KR (1972) Electron transport system of *Nitrosomonas*: isolation of a membrane-envelope fraction. J Bacteriol 110:430–438

Hooper AB, Nason A (1965) Characterization of hydroxylamine-cytochrome *c* reductase from the chemoautotrophs *Nitrosomonas europaea* and *Nitrosocystis oceanus*. J Biol Chem 240: 4044–4057

Hooper AB, Terry KR (1973) Specific inhibition of ammonia oxidation in *Nitrosomonas*. J Bacteriol 115:480–485

Hooper AB, Terry KR (1979) Hydroxylamine oxidoreductase of *Nitrosomonas*. Production of nitric oxide from hydroxylamine. Biochim Biophys Acta 571:12–20

Hooper AB, Terry KR, Maxwell PC (1977) Hydroxylamine oxidoreductase of *Nitrosomonas*. Oxidation of diethyldithiocarbamate concomitant with stimulation of nitrite synthesis. Biochim Biophys Acta 462:141–152

Houchis JP, Hind G (1983) Flash spectroscopic characterization of photosynthetic electron transport in isolated heterocysts. Arch Biochem Biophys 224:272–282

Hu S-I, Pezacka E, Wood HG (1984) Acetate synthesis from carbon monoxide by *Clostridium thermoaceticum*. Purification of corrinoid protein. J Biol Chem 259:8892–8897

Huber R, Kristjansson JK, Stetter KO (1987) *Pyrobaculum* gen. nov., a new genus of neutrophilic, rod-shaped archaebacteria from continental solfataras growing optimally at 100°C. Arch Microbiol 149:95–101

Hyman MR, Arp DJ (1995) Effects of ammonia on the de novo synthesis of polypeptides in cells of *Nitrosomonas europaea* denied ammonia as an energy source. J Bacteriol 177:4974–4979

Hyman MR, Russel SA, Ely RL, Williamson KJ, Arp DJ (1995) Inhibition, inactivation, and recovery of ammonia-oxidizing activity in cometabolism of trichloroethylene by *Nitrosomonas europaea*. Appl Environ Microbiol 61:1480–1487

Hyman MR, Wood PM (1983) Methane oxidation by *Nitrosomonas europaea*. Biochem J 212:31–37

Hyman MR, Wood PM (1985) Suicidal inactivation and labelling of ammonia monooxygenase by acetylene. Biochem J 227:719–725

Igarashi N, Noriyama H, Fujiwara T, Fukumori Y, Tanaka N (1997) The 2Å structure of hydroxylamine oxidoreductase from a nitrifying chemoautotrophic bacterium, *Nitrosomonas europaea*. Nat Struct Biol 4:276–284

Imai K (1984) The autotrophic bacteria (in Japanese). Kagaku Doujin, Kyoto

Imhoff JF (1983) *Rhodopseudomonas marina* sp. nov., a new marine phototrophic purple bacterium. Syst Appl Microbiol 4:512–521

Inatomi K-J (1986) Characterization and purification of the membrane-bound ATPase of the archaebacterium *Methanosarcina barkeri*. J Bacteriol 167:837–841

Ingledew WJ (1982) *Thiobacillus ferrooxidans*. The bioenergetics of an acidophilic chemolithoautotroph. Biochim Biophys Acta 683:89–117

Ingledew WJ, Houston A (1986) The organization of the respiratory chain of *Thiobacillus ferrooxidans*. Biotechnol Appl Biochem 8:242–248

Inoue C, Takeshima T, Matsumoto N (1995) Recovery of gold from the fine ore containing iron sulfide by the bacterial leaching. Resources. Materials '95 (Suruga) Application of microorganisms to the exploit of resources (in Japanese). pp 15–18

Ishii M, Miyake T, Satoh T, Sugiyama H, Oshima Y, Kodama T, Igarashi Y (1997) Autotrophic carbon dioxide fixation in *Acidianus brierleyi*. Arch Microbiol 166:368–371

Ishikawa Y (1991) Kuroko: searching for world-famous Japanese mineral resources (in Japanese). Kyoritsu, Tokyo

Ishimoto M (1959) Sulfate reduction in cell-free extracts of *Desulfovibrio*. J Biochem 46:105–106

Ishimoto M, Koyama J, Nagai Y (1954) Biochemical studies on sulfate-reducing bacteria IV. The cytochrome system of sulfate-reducing bacteria. J Biochem 41:763–770

Ivanovsky RN, Krasilnikova EN, Fal YI (1993) A pathway of the autotrophic $CO_2$ fixation in *Chloroflexus aurantiacus*. Arch Microbiol 159:257–264

Iverson TM, Arciero DM, Hooper AB, Rees DC (2001) High solution structures of the oxidized and reduced states of cytochrome *c*-554 from *Nitrosomonas europaea*. J Biol Inorg Chem 6:390–397

Iwahori K, Kamimura K, Sugio T (1998) Isolation and some properties of cytochrome *c* oxidase purified from a bisulfite ion resistant *Thiobacillus ferrooxidans* strain, OK1–50 Biosci Biotechnol Biochem 62:1081–1086

Iwasaki H, Matsubara T (1972) A nitrite reductase from *Achromobacter cycloclastes*. J Biochem 71:645–652

Iwata S, Ostermeier C, Ludwig B, Michel H (1995) Structure at 2.8Å resolution of cytochrome *c* oxidase from *Paracoccus denitrificans*. Nature 376:660–669

Jannasch HW, Nelson DC, Wirsen CO (1989) Massive natural occurrence of unusually large bacteria (*Beggiatoa* sp.) at a hydrothermal deep-sea vent site. Nature 342:834–836

Jetten MSM, de Bruijn P, Kuenen JG (1997) Hydroxylamine metabolism in *Pseudomonas* PB16: involvement of a novel hydroxylamine oxidoreductase. Antonie van Leeuwenhoek. 71:69–74

Jetten MSM, Strous M, van de Pas-Schoonen KT, Schalk J, van Dongen UGJM, van de Graaf AA, Logemann S, Muyzer G, van Loosdrecht MCM, Kuenen JG (1999) The anaerobic oxidation of ammonium. FEMS Microbiol Rev 22:421–437

Johnson MS, Zhulin IB, Gapuzan M-ER, Tayler BL (1997) Oxygen-dependent growth of the obligate anaerobe *Desulfovibrio vulgaris* Hildenborough. J Bacteriol 179:5598–5601

Jørgensen BB, Gallardo VA (1999) *Thioploca* spp: filamentous sulfur bacteria with nitrate vacuoles. FEMS Microbiol Ecol 28:301–313

Juliette LY, Hyman MR, Arp DJ (1993) Mechanism-based inactivation of ammonia monooxygenase in *Nitrosomonas europaea* by allylsulfide. Appl Environ Microbiol 59:3728–3735

Kadenbach B (1983) Structure and evolution of the "Atmungsferment" cytochrome *c* oxidase. Angew Chem Int ed Engl 22:275–283

Kai M, Yano T, Fukumori Y, Yamanaka T (1989) Cytochrome *c* oxidase of acidophilic iron-oxidizing bacterium, *Thiobacillus ferrooxidans*, functions at pH 3.5. Biochem Biophys Res Commun 160:839–843

Kai M, Yano T, Tamegai H, Fukumori Y, Yamanaka T (1992) *Thiobacillus ferrooxidans* cytochrome *c* oxidase: purification, and molecular and enzymatic features. J Biochem 112:816–821

Kamimura K, Fujii S, Sugio T (2001) Purification and some properties of ubiquinol oxidase from obligately chemoautotrophic iron-oxidizing bacterium, *Thiobacillus ferrooxidans* NASF-1. Biosci Biotechnol Biochem 65:63–71

Kamiyoshi H, Tabata M (1991) On the substances which stimulate growth of ammonia oxidizing bacterium *Nitrosomonas europaea* (in Japanese). Nihon Nogeikagaku Kaishi (J Jpn Agr Chem Soc) 65:442 (Abstracts of papers at the annual meeting)

Kappler U, Bennett B, Rethmeier J, Schwarz G, Deutzmann R, McEwan AG, Dahl C (2000) Sulfite cytochrome *c* oxidoreductase from *Thiobacillus novellus*. Purification, characterization, and molecular biology of a heterodimeric member of sulfite oxidase family. J Biol Chem 275:13202–13212

Kappler U, Fridrich CG, Trüper HG, Dahl C (2001) Evidence for two pathways of thiosulfate oxidation in *Starkeya novella* (formerly *Thiobacillus novellus*). Arch Microbiol 175:102–111

Karlin KD (1993) Metalloenzymes, structural motives and inorganic models. Science 261:701–708

Kashefi K, Lovley DR (2000) Reduction of Fe(III), Mn(IV), and toxic metals at 100°C by *Pyrobaculum islandicum.* Appl Environ Microbiol 66:1050–1056

Kashefi K, Tor JM, Holmes DC, Gaw Van Praagh CV, Reysenbach AL, Lovley DR (2002) *Geoglobus ahangari* gen. nov., sp. nov., a novel hyperthermophilic archaeon capable of oxidizing organic acids and growing autotrophically on hydrogen with Fe(III) serving as the sole electron acceptor. Int J Syst Evol Microbiol 52:719–728

Kashefi K, Tor JM, Navin KP, Lovley DR (2001) Reductive precipitation of gold by dissimilatory Fe(III)-reducing *Bacteria* and *Archaea.* Appl Environ Microbiol 67:3275–3279

Keilin D (1966) The history of cell respiration and cytochrome. Cambridge University Press, London

Keilin D, Hartree EF (1939) Cytochrome and cytochrome oxidase. Proc R Soc B 127:167–191

Kelly DP (1989) Oxidation of sulphur compounds. In: Cole JA, Ferguson SJ (eds) Nitrogen and sulphur cycles. Cambridge University Press, Cambridge, pp 65–98

Kelly DP, Lu W-P, Poole RK (1993) Cytochromes in *Thiobacillus tepidarius* and the respiratory chain involved in the oxidation of thiosulfate and tetrathionate. Arch Microbiol 160:87–95

Kelly MJ, Poole RK, Yates MG, Kennedy C (1990) Cloning and mutagenesis of genes encoding the cytochrome *bd* terminal oxidase complex in *Azotobacter vinelandii*: mutants deficient in cytochrome *d* complex were unable to fix nitrogen in air. J Bacteriol 172:6010–6019

Kepkay PE, Nealson KH (1987) Growth of manganese oxidizing *Pseudomonas* sp. in continuous culture. Arch Microbiol 148:63–67

Kessler DL, Rajagopalan KV (1972) Purification and properties of sulfite oxidase from chicken liver. J Biol Chem 247:6566–6573

Ketchum PA, Sanders HK, Grydes JW, Nason A (1969) Characterization of cytochrome *c* from *Nitrobacter agilis.* Biochim Biophys Acta 189:360–365

Kim J, Rees DC (1992) Structural models for the metal centers in the nitrogenase molybdenum-iron protein. Science 257:1677–1682

Kimura M, Miura Y, Watanabe A, Katoh T, Haraguchi H (1991) Methane emission from paddy field (part 1). Effect of fertilization, growth stage and midsummer drainage: pot experiment. Environ Sci 4:265–271

Kinsel NA, Umbreit WW (1964) Method for electrophoresis of culture medium to increase growth of the sulfur-oxidizing iron bacterium *Ferrobacillus sulfooxidans.* J Bacteriol 87:1243–1246

Kitagawa T, Ozaki Y, Teraoka J, Kyogoku Y, Yamanaka T (1977) The pH dependence of the resonance Raman spectra and structural alterations at heme moieties of various *c*-type cytochromes. Biochim Biophys Acta 494:100–114

Kitamura M, Mizugai K, Taguchi M, Akutsu H, Kumagai I, Nakaya T (1995) A gene encoding a cytochrome *c* oxidase-like protein is located closely to the cytochrome *c*-553 gene in the anaerobic bacterium, *Desulfovibrio vulgaris* (Miyazaki F). Microbiol Immunol 39:75–80

Kletzin A (1989) Coupled enzymatic production of sulfite, thiosulfite, and hydrogen sulfide from sulfur: purification and properties of a sulfur oxygenase reductase from the facultative anaerobic archaebacterium *Desulfurolobus ambivalens*. J Bacteriol 171:1638–1643

Kobayashi K, Takahashi E, Ishimoto M (1972) Biochemical studies on sulfate reducing bacteria IX. Purification and some properties of sulfite reductase, desulfoviridin. J Biochem 72:879–887

Koga Y (1988) Archaebacteria (in Japanese). Tokyo University Press, Tokyo

Koga Y (2004) Diversification of bacteria, archaea, and eucarya—a hypothesis based on the enantiomeric glycerophospholipids (in Japanese). Seikagaku 76:349–358

Kokufuta E, Shimohashi M, Nakamura I (1988) Simultaneously occurring nitrification and denitrification under oxygen gradient by poly electrolyte complex-coimmobilized *Nitrosomonas europaea* and *Paracoccus denitrificans* cells. Biotech Bioeng 31:382–384

Koops H-P, Bottcher B, Moller UC, Pommerening-Roser A, Stehr G (1991) Classification of eight new species of ammonia-oxidizing bacteria: *Nitrosomonas communis* sp nov., *Nitrosomonas ureae* sp. nov., *Nitrosomonas aestuarii*, sp.nov., *Nitrosomonas marina* sp. nov., *Nitrosomonas nitrosa*, sp. nov., *Nitrosomonas eutropha*, sp. nov., *Nitrosomonaas oligotropha* sp. nov. and *Nitrosomonas halophila* sp. nov. J Gen Microbiol 137:1689–1699

Kreil G (1963) Über die Artspezifität con Cytochrome *c*. Vergleigh der Aminosäuresequenz des Thunfisch-Cytochroms *c* mit der des Pferde-Cytochroms *c*. Z Physiol Chemie 334:154–166

Kreil G (1965) Die c-terminale Aminosäuresequenz des Thunfisch-Cytochroms *c*. Z Physiol Chemie 340:86–87

Krümmel A, Harms H (1982) Effect of organic matter on growth and cell yield of ammonia oxidizing bacteria. Arch Microbiol 133:50–54

Kubo H (1939) Über das Hämoprotein aus den Wurzelknollchen von Leguminosen. Acta Phytochim 11:195–200

Kucera S, Wolfe RS (1957) A selective enrichment method for *Gallionella ferruginea*. J Bacteriol 74:344–346

Kuenen JG (1989) Colorless sulfur bacteria. In: Stanley JT, Bryant MP, Pfennig N, Holt JG (eds) Bergey's manual of systematic bacteriology vol 3. Williams & Wilkins, Baltimore, pp 1834–1871

Kühn W, Fiebig K, Walther R, Gottschalk G (1979) Presence of cytochrome *b*-559 in *Methanosarcina barkeri*. FEBS Lett 105:1271–1274

Kühn W, Fiebig K, Hippe H, Mah RA, Huser BA, Gottschalk G (1983) Distribution of cytochromes in methanogenic bacteria. FEMS Microbiol Lett 20:407–410

Kühn W, Gottschalk G (1983) Characterization of the cytochromes occurring in *Methanosarcina* species. Eur J Biochem 135:89–94

Kumazawa Y, Fujiwara T, Fukumori Y, Koga Y, Yamanaka T (1994) Cytochrome *bc* purified from the methanogen *Methanosarcina barkeri*. Curr Microbiol 29:53–56

Kurihara Y (1999) Exploitation of anti-bacterial concrete which guards concrete from its corrosion (in Japanese). Cement Concrete 633, 44–49

Kurokawa M, Fukumori Y, Yamanaka T (1985) A hydroxylamine-cytochrome *c* reductase occurs in the heterotrophic nitrifier, *Arthrobacter globiformis*. Plant Cell Physiol 26:1439–1442

Kurokawa T, Fukumori Y, Yamanaka T (1987) Purification of a flavoprotein having NADPH-cytochrome *c* reductase and transhydrogenase activities from *Nitrobacter winogradskyi* and its molecular and enzymatic properties. Arch Microbiol 148:95–99

Kurokawa T, Fukumori Y, Yamanaka T (1989) *Nitrobacter winogradskyi* cytochrome *b*-559:a nonhaem iron-containing cytochrome related to bacterioferritin. Biochim Biophys Acta 976: 135–139

Kuronen T, Ellfolk N (1972) A new purification procedure and molecular properties of *Pseudomonas* cytochrome oxidase. Biochim Biophys Acta 275:308–318

Kusano T, Takeshima T, Sugawara K, Inoue C, Shiratori T, Yano T, Fukumori Y, Yamanaka T (1992) Molecular cloning of the gene encoding *Thiobacillus ferrooxidans* Fe(II) oxidase. High homology of the gene product with HiPIP. J Biol Chem 267:11242–11247

Lampreia J, Moura L, Teixeria M, Peck HD Jr, LeGall J, Huynth BH, Moura JJG (1990) The active centers of adenylylsulfate reductase from *Desulfovibrio gigas*. Characterization and spectroscopic studies. Eur J Biochem 188:653–664

Lang E, Jagnow G (1986) Fungi of a forest soil nitrifying at low pH values. FEMS Microbiol Ecol 38:257–265

Langworthy TA, Holzer G, Zeikus JG, Tornabene TG (1983) Iso- and anteiso-branched glycerol diethers of the thermophilic anaerobe *Thermodesulfotobacterium commune*. Syst Appl Microbiol 4:1–17

Laska S, Lottspeich F, Kletzin A (2003) Membrane-bound hydrogenase and sulfur reductase of hyperthermophilic and acidophilic archaeon *Acidianus ambivalens*. Microbiology 149:2357–2371

Lauraeus M, Haltia T, Saraste M, Wikström M (1991) *Bacillus subtilis* expresses two kinds of haem A-containing terminal oxidases. Eur J Biochem 197:695–705

Lazaroff N (1963) Sulfate requirement for iron oxidation by *Thiobacillus ferrooxidans*. J Bacteriol 85:78–83

Lederer F, Simon AM, Verdière J (1972) *Saccharomyces cerevisiae* iso-cytochromes *c*: revision of the amino acid sequence between the cysteine residues. Biochem Biophys Res Commun 47:55–58

Leduc LG, Ferroni GD (1994) The chemolithotrophic bacterium *Thiobacillus ferrooxidans*. FEMS Microbiol Rev 14:103–120

Lee J-P, Peck HD Jr (1971) Purification of the enzyme reducing bisulfite to trithionate from *Desulfovibrio gigas* and its identification as desulfoviridin. Biochem Biophys Res Commun 45:583–589

Lees H (1952) Hydroxylamine as an intermediate in nitrification. Nature 169:156–157

Lees H, Simpson JR (1957) The biochemistry of the nitrifying organisms 5. Nitrite oxidation by *Nitrobacter*. Biochem J 65:297–305

LeGall J, Peck HD Jr (1987) Amino-terminal amino acid sequences of electron transfer proteins from Gram-negative bacteria as indicators of their cellular localization: the sulfate-reducing bacteria. FEMS Microbiol Rev 46:35–40

Leigh JA, Rinehart KL Jr, Wolfe RS (1985) Methanofuran (carbon dioxide reduction factor), a formyl carrier in methane production from carbon dioxide in *Methanobacterium*. Biochemistry 24:995–999

Lin D, Kakizono T, Nishio N, Nagai S (1990) Enhanced cytochrome formation and stimulated methanogenesis rate by the increased ferrous concentrations in *Methanosarcina barkeri* culture. FEMS Microbiol Lett 68:89–92

Liu C-L, Peck HD Jr (1981a) Comparative bioenergetics of sulfate reduction in *Desulfovibrio* and *Desulfotomaculum* spp J Bacteriol 145:966–973

Liu M-C, Costa C, Coutinho IB, Moura I, Xavier AV, LeGall J (1988) Cytochrome components of nitrate- and sulfate-respiring *Desulfovibrio desulfuricans* ATCC 27774. J Bacteriol 170:5545–5551

Liu M-C, Peck HD Jr (1981b) The isolation of hexaheme cytochrome from *Desulfovibrio desulfuricans* and identification as a new type of nitrite reductase. J Biol Chem 256:13159–13164

Loufti MG, Guerlesquin F, Bianco P, Haladjian J, Bruschi M (1989) Comparative studies of polyhemic cytochromes *c* isolated from *Desulfovibrio vulgaris* (Hildenborough) and *Desulfovibrio desulfuricans* (Norway). Biochem Biophys Res Commun 159:670–676

Lovley DR, Giovannoni SJ, White DC, Champine JE, Phillips EJ, Gorby YA, Goodwin S (1993) *Geobacter metallireducens* gen. nov. sp. nov., a microorganism capable of coupling the complete oxidation of organic compounds to reduction of iron and other metals. Arch Microbiol 159:336–344

Lowe DR (1994) Early environments: constraints and opportunities for early evolution. In: Bengtson S (ed) Early life on Earth. Nobel symposium No. 84. Columbia University Press, New York, pp 24–35

Lu W-P, Kelly DP (1983a) Rhodanese: an enzyme not necessary for thiosulphate oxidation by *Thiobacillus* A2. FEMS Microbiol Lett 18:289–292

Lu W-P, Kelly DP (1983b) Purification and some properties of two principal enzymes of the thiosulphate-oxidizing multi-enzyme system from *Thiobacillus* A2. J Gen Microbiol 129: 3549–3564

Lu W-P, Kelly DP (1984a) Properties and role of sulphite:cytochrome *c* oxidoreductase purified from *Thiobacillus versutus* (A2). J Gen Microbiol 130:1683–1692

Lu W-P, Kelly DP (1984b) Purification and characterization of two essential cytochromes of the thiosulphate-oxidizing multi-enzyme system from *Thiobacillus* A2 (*Thiobacillus versutus*) Biochim Biophys Acta 765:106–117

Lu W-P, Kelly DP (1988a) Cellular location and partial purification of the "thiosulphate-oxidizing enzyme" and "trithionate hydroxylase" from *Thiobacillus tepidarius.* J Gen Microbiol 134:877–885

Lu W-P, Kelly DP (1988b) Respiration-driven proton translocation in *Thiobacillus versutus* and the role of the periplasmic thiosulphate-oxidizing enzyme system. Arch Microbiol 149: 297–302

Lübben M, Kolmerer B, Saraste M (1992) On archaebacterial terminal oxidase combines core structures of two mitochondrial respiratory complexes. EMBO J 11:805–812

Lundgren DG, Silver M (1980) Ore leaching by bacteria. Annu Rev Microbiol 34:263–283

Lyric RM, Suzuki I (1970a) Enzyme involved in the metabolism of thiosulfate by *Thiobacillus thioparus* III. Properties of thiosulfate-oxidizing enzyme and proposed pathway of thiosulfate oxidation. Can J Biochem 48:355–363

Lyric RM, Suzuki I (1970b) Enzymes involved in the metabolism of thiosulfate by *Thiobacillus thioparus* I. Survey of enzymes and properties of sulfite:cytochrome *c* oxidoreductase. Can J Biochem 48:334–343

Lyric RM, Suzuki I (1970c) Kinetic studies of sulfite:cytochrome *c* oxidoreductase, thiosulfate-oxidizing enzyme, and adenosine-5′-phosphosulfate reductase from *Thiobacillus thioparus.* Can J Microbiol 48:594–603

MacGregor RA (1964) Recovery of $U_3O_8$ by underground leaching. Transactions 69:162–166

Magnuson TS, Hodges-Myerson AL, Lovley DR (2000) Characterization of membrane-bound NADH-dependent $Fe^{3+}$ reductase from the dissimilatory $Fe^{3+}$-reducing bacterium *Geobacter sulfurreducens.* FEMS Microbiol Lett 185:205–211

Malashenko YuR, Sokolov IG, Romanovskaya VA, Shkurko Yu B (1979) Element of lithotrophic metabolism in the obligate methylotroph *Methylococcus thermophilus.* Mikrobiologia (translated) 48:592–598

Margoliash E, Smith EL, Kreil G, Tuppy H (1961) Amino acid sequence of horse-heart cytochrome *c.* Nature 192:1121–1127

Martin PAW, Dugan PR, Tuovinen OH (1983) Uranium resistance of *Thiobacillus ferrooxidans.* Eur J Appl Microbiol Biotechnol 18:392–395

Matsuba D, Takazaki H, Sato Y, Takahashi R, Tokuyama T, Wakabayashi K (2003) Susceptibility of ammonia-oxidizing bacteria to nitrification inhibitors. Z Naturforsch 58c:282–287

Matsubara H, Smith EL (1963) Human heart cytochrome *c.* J Biol Chem 238:2732–2753

Matsumoto N, Nakasono S, Ohmura N, Saiki H (1999) Extension of logarithmic growth of *Thiobacillus ferrooxidans* by potential controlled electrochemical reduction of Fe(III). Biotechnol Bioeng 64:716–721

Matsushita K, Shinagawa E, Adachi O, Ameyama M (1982) *o*-Type cytochrome oxidase in the membrane of aerobically grown *Pseudomonas aeruginosa.* FEBS Lett 139:285–288

Matsushita K, Shinagawa E, Adachi O, Ameyama M (1990) Cytochrome $a_1$ of *Acetobacter aceti* is a cytochrome *ba* functioning as ubiquinol oxidase. Proc Natl Acad Sci USA 87:9863–9867

Mayer F, Jussofie A, Salzmann M, Lübben M, Rhode M, Gottschalk G (1987) Immunoelectron microscopic demonstration of ATPase on the cytoplasmic membrane of the methanogenic bacterium strain Gö 1. J Bacteriol 169:2307–2309

Maymo-Gatell X, Chien Y-T, Gossett JM, Zinder SH (1997) Isolation of a bacterium that reductively dehalogenate tetrachloroethene to ethene. Science 276:1568–1571

McTavish H, Arciero DM, Hooper AB (1995) Interaction with membranes of cytochrome *c*-554 from *Nitrosomonas europaea.* Arch Biochem Biophys 324:53–58

McTavish H, Fuchs JA, Hooper AB (1993) Sequence of the gene coding for ammonia monooxygenase in *Nitrosomonas europaea.* J Bacteriol 175:2436–2444

Meincke M, Bock E, Kastrau D, Kroneck PMH (1992) Nitrite oxidoreductase from *Nitrobacter hamburgensis*: redox centers and their catalytic role. Arch Microbiol 158:127–131

Meulenberg R, Pronk JT, Hazeu W, van Dijken JP, Frank J, Bos P, Kuenen JG (1993) Purification and partial characterization of thiosulphate dehydrogenase from *Thiobacillus acidophilus.* J Gen Microbiol 139:2033–2039

Meyer TE, Kamen MD (1982) New perspectives on *c*-type cytochromes. Adv Protein Chem 35:105–212

Miller DJ, Wood PM (1983c) The soluble oxidase of *Nitrosomonas europaea.* J Gen Microbiol 129:1645–1650

Miller DJ, Wood PM, Nicholas DJD (1984) Further characterization of cytochrome P-460 in *Nitrosomonas europaea.* J Gen Microbiol 130:3049–3054

Miller SL (1953) A production of amino acids under possible primitive Earth conditions. Science 117:528–529

Mogi T, Tsubaki M, Hori H, Miyoshi H, Nakamura H, Anraku Y (1998) Two terminal quinol oxidase families in *Escherichia coli*: variations on molecular machinery for dioxygen reduction. J Biochem Mol Biol Biophys 2:79–110

Moir JWB, Crossman LC, Spiro S, Richardson DJ (1996) Purification of ammonia monooxygenase from *Paracoccus denitrificans.* FEBS Lett 387:71–74

Moore GR, Pettigrew GW (1990) Cytochromes *c*: Evolutionary, structural and physicochemical aspects. Springer, Berlin

Mori T, Nonaka T, Tazaki K, Koga M, Hikosaka Y, Noda S (1992) Interactions of nutrients, moisture and pH on microbial corrosion on concrete sewer pipes. Wat Res 26:29–37

Moriarty DJW, Nicholas DJD (1970) Electron transfer during sulphide and sulphite oxidation by *Thiobacillus concretivorus.* Biochim Biophys Acta 216:130–138

Morikawa M, Imanaka T (1993) Isolation of a new mixotrophic bacterium which can fix $CO_2$ and assimilate aliphatic and aromatic hydrocarbons anaerobically. J Ferment Bioeng 76: 280–283

Morton RK (1958) The cytochromes. Rev Pure Appl Chem 8:161–220

Mucha H, Keller E, Weber H, Lingens F, Trosch W (1985) Sirohydrochlorin, a precursor of factor $F_{430}$ biosynthesis in *Methanobacterium thermoautotrophicum.* FEBS Lett 190:169–171

Mulder EG (1989) Genus *Leptothrix* Kutzing 1943. In: Stanley JT, Bryant MP, Pfennig N, Holt JG (eds) Bergey's manual of systematic bacteriology vol 3. Williams & Wilkins, Baltimore, pp 1998–2003

Mulkidjanian AY, Koonin EV, Makarova KS, Mekhedov SL, Sorokin A, Wolf YI, Dufresne A, Partensky F, Burd H, Kaznadzey D, Haselkorn R, Galperin MY (2006) The cyanobacterial genome core and the origin of photosynthesis. Proc Natl Acad Sci USA 103:13126–13131

Murphy MJ, Siegel LM (1973) Siroheme and sirohydrochlorin. The basis for a new type of porphyrin-related prosthetic group common to both assimilatory and dissimilatory sulfite reductase. J Biol Chem 248:6911–6919

Murphy MJ, Siegel LM, Kamin H (1973) Reduced nicotinamide adenine dinucleotide phosphate-sulfite reductase of enterobacteria II. Identification of a new class of heme prosthetic group: an iron-tetrahydroporphyrin (isobacteriochlorin type) with eight carboxylic acid groups. J Biol Chem 248:2801–2814

Nagata Y, Yamanaka T, Okunuki K (1970) Amino acid composition and *N*-terminus of *Pseudomonas* cytochrome oxidase (= *Pseudomonas aeruginosa* nitrite reductase). Biochim Biophys Acta 221:668–671

Nakagawa A, Higuchi Y, Yasuoka N, Katsube Y, Yagi T (1990) S-class cytochromes *c* have a variety of folding patterns: structure of cytochrome *c*-553 from *Desulfovibrio vulgaris* determined by the multi-wavelength anomalous dispersion method. J Biochem 108:701–703

Nakamura K, Nakamura M, Yoshikawa H, Amano Y (2001) Purification and properties of thiosulfate dehydrogenase from *Acidithiobacillus thiooxidans* JCM 7814. Biosci Biotechnol Biochem 165:102–108

Nakamura K, Yoshikawa H, Okubo S, Kurosawa H, Amano Y (1995) Purification and properties of membrane-bound sulfite dehydrogenase from *Thiobacillus thiooxidans* JCM 7814. Biosci Biotechnol Biochem 59:11–15

Nakasono S, Matsumoto N, Saiki H (1997) Electrochemical cultivation of *Thiobacillus ferrooxidans* by potential control. Bioelectrochem Bioenerg 43:61–66

Narita K, Titani K (1969) The complete amino acid sequence in baker's yeast cytochrome *c*. J Biochem 65:259–267

Nell KM, Rinehart KL Jr, Tanner RS, Wolfe RS (1986) Structure of component B (7-mercaptoheptanoylthreonine phosphate) of the methylcoenzyme M methylreductase system of *Methanobacterium thermoautotrophicum*. Proc Natl Acad Sci USA 83:4238–4242

Nevin KP, Lovley DR (2000) Lack of production of electron shuttling compound or solubilization of Fe(III) during reduction of insoluble Fe(III) oxide by *Geobacter metallireducens*. Appl Environ Microbiol 66:2248–2251

Nisbet EG, Cann R, Van Dover CL (1995) Origins of photosynthesis. Nature 373:479–480

Nishio M (1981) The harm of nitric oxide evolved from soil in greenhouse. In: Furusaka C (ed) Microorganisms in soil (in Japanese). Hakuyu-sha, Tokyo, pp 109–111

Nishio T, Yoshimura T, Chiba K, Inoue Z (1994) Effects of organic acids on heterotrophic nitrification by *Alcaligenes faecalis* OKK17. Biosci Biotech Biochem 58:1574–1578

Nogami Y, Maeda T, Negishi A, Sugio T (1997) Inhibition of sulfur oxidizing activity by nickel ion in *Thiobacillus ferrooxidans* NB1–3 isolated from the corroded concrete. Biosci Biotech Biochem 61:1373–1375

Nomoto T, Fukumori Y, Yamanaka T (1993) Membrane-bound cytochrome *c* is an alternative electron donor for cytochrome $aa_3$ in *Nitrobacter winogradskyi*. J Bacteriol 175:4400–4404

Numata M (1989) Studies on hydroxylamine oxidation in *Nitrosomonas europaea*. Dissertation for Masters degree, Tokyo Institute of Technology

Numata M, Saito T, Yamazaki T, Fukumori Y, Yamanaka T (1990) Cytochrome P-460 of *Nitrosomonas europaea*: further purification and further characterization. J Biochem 108:1016–1021

Numata M, Yamazaki T, Fukumori Y, Yamanaka T (1989) Some properties of *Nitrosomonas europaea* cytochrome *c* oxidase ($aa_3$-type) which lacks CuA. J Biochem 105:245–248

Numoto N, Nakagawa T, Kita A, Sasayama Y, Fukumori Y, Miki K (2005) Structure of an extracellular giant hemoglobin of the gutless beard worm *Oligobranchia mashikoi*. Proc Natl Acad Sci USA 102:14521–14526

Nunzi F, Woudstra M, Compese D, Bonicel J, Morin D, Bruschi M (1993) Amino-acid sequence of rusticyanin from *Thiobacillus ferrooxidans* and its comparison with other blue copper proteins. Biochim Biophys Acta 1162:28–34

O'Brian MR, Kirshbom PM, Maier RJ (1987) Bacterial heme synthesis is required for expression of the leghemoglobin protein but not the apoprotein in soybean root nodules. Proc Natl Acad Sci USA 84:8390–8393

Ogata M, Arihira K, Yagi T (1981) D-Lactate dehydrogenase of *Desulfovibrio vulgaris*. J Biochem 89:1423–1431

Ogata M, Kondo S, Okawara N, Yagi T (1988) Purification and characterization of ferredoxin from *Desulfovibrio vulgaris* Miyazaki. J Biochem 103:121–125

Oh JK, Suzuki I (1977a) Isolation and characterization of a membrane-associated thiosulphate-oxidizing system of *Thiobacillus novellus*. J Gen Microbiol 99:397–412

Oh JK, Suzuki I (1977b) Resolution of membrane-associated thiosulphate-oxidizing complex of *Thiobacillus novellus*. J Gen Microbiol 99:413–423

Ohmura N, Matsumoto N, Sasaki K, Saiki H (2002) Electrochemical regeneration of Fe(III) to support growth on anaerobic iron respiration. Appl Environ Microbiol 68:405–407

Ono Y, Enokiya A, Masuko D, Shoji K, Yamanaka T (1999) Pyruvic oxime dioxygenase from the heterotrophic nitrifier *Alcaligenes faecalis*: purification, and molecular and enzymatic properties. Plant Cell Physiol 40:47–52

Ono Y, Makino N, Hoshino Y, Shoji K, Yamanaka T (1996) An iron dioxygenase from *Alcaligenes faecalis* catalyzing the oxidation of pyruvic oxime to nitrite. FEMS Microbiol Lett 139:103–108

Onysko SJ, Kleinmann RLP, Erickson PM (1984) Ferrous iron oxidation by *Thiobacillus ferrooxidans*: inhibition with benzoic acid, sorbic acid, and sodium lauryl sulfate. Appl Environ Microbiol 48:229–231

Orii Y, Webster DA (1986) Photodissociation of oxygenated cytochrome *o* (s) (*Vitreoscilla*) and kinetic studies of reassociation. J Biol Chem 261:3544–3547

Oro J (1994) Early chemical stages in the origin of life. In: Bengtson S (ed) Early life on Earth. Nobel symposium No. 84. Columbia University Press, New York, pp 48–59

Otsuka S, Yamanaka T (eds) (1988) Metalloproteins: chemical properties and biological effects. Kodansha, Tokyo/Elsevier, Amsterdam

Otte S, Schalk J, Kuenen JG, Jetten MSM (1999) Hydroxylamine oxidation and subsequent nitrous oxide production by the heterotrophic ammonia oxidizer *Alcaligenes faecalis.* Appl Microbiol Biotechnol 51:255–261

Oyama T, Chigira M, Ohmura N, Watanabe Y (1998) Heave of house foundation by the chemical weathering of mudstone (in Japanese). Oyo Chisitsu (J Jpn Soc Eng Geol) 39:261–272

Parker CD (1945) The corrosion of concrete I. The isolation of a species of bacterium associated with the corrosion of concrete exposed to atmosphere containing hydrogen sulfide. Aust J Exp Med Sci 23:81–90

Paula MA, Reis VM, Dobereiner J (1991) Interactions of *Glomus clarum* with *Acetobacter diazotrophicus* in infection of sweet potato (*Impomoea batatas*), sugarcane (*Saccharum* spp), and sweet sorghum (*Sorghum vulgare*). Biol Fertil Soils 11:111–115

Pearson AR, Elmore BO, Yang C, Ferrara JD, Hooper AB, Wilmot CM (2007) The crystal structure of cytochrome P460 of *Nitrosomonas europaea* reveals a novel cytochrome fold and heme-protein cross-link. Biochemistry 46:8340–8349

Peck HD Jr (1959) The ATP-dependent reduction of sulfate with hydrogen in extracts of *Desulfovibrio desulfuricans.* Proc Natl Acad Sci USA 45:701–708

Peck HD Jr (1960) Evidence for oxidative phosphorylation during the reduction of sulfite with hydrogen by *Desulfovibrio desulfuricans.* J Biol Chem 235:2734–2738

Peck HD Jr (1966) Phosphorylation coupled with electron transfer in extracts of the sulfate-reducing bacterium *Desulfovibrio gigas*. Biochem Biophys Res Commun 22:112–118

Peer CW, Painter MH, Rasche ME, Ferry JG (1994) Characterization of a CO:heterodisulfide oxidoreductase system from acetate-grown *Methanosarcina thermophila.* J Bacteriol 176: 6974–6979

Pereira IAC, Pacheco I, Liu M-Y, LeGall J, Xavier AV, Teixeira M (1997) Multiheme cytochromes from the sulfur-reducing bacterium *Desulfuromonas acetoxidans.* Eur J Biochem 248:323–328

Peterson RB, Wolk CP (1978) High recovery of nitrogenase activity and of $^{55}$Fe-labelled nitrogenase in heterocysts isolated from *Anabaena variabilis.* Proc Natl Acad Sci USA 75: 6271–6275

Pettigrew GW, Moore GR (1987) Cytochromes *c*: biological aspects. Springer, Berlin

Pezacka E, Wood HG (1986) The autotrophic pathway of acetogenic bacteria: role of CO dehydrogenase disulfide reductase. J Biol Chem 261:1609–1615

Pfaltz A, Jaun B, Fassler A, Eschenmoser A, Jaenchen,R, Gillee HH, Diekert G, Thauer RK (1982) Zur Kenntnis des Faktors $F_{430}$ aus methanogen Bakterien: Structur des porphinoiden Ligandsystems. Helv Chim Acta 65:828–826

Pfennig N, Biebl H (1976) *Desulfuromonas acetoxidans* gen. nov. and sp. nov., a new anaerobic, sulfur-reducing, acetate-oxidizing bacterium. Arch Microbiol 110:3–12

Poole RK, Hill S (1997) Respiratory protection of nitrogenase activity in *Azotobacter vinelandii*—roles of the terminal oxidases. Biosci Rep 17:303–317

Postgate JR (1954) Presence of cytochrome in obligate anaerobe. Biochem J 56:xi–xii
Postgate JR (1956) cytochrome $c_3$ and desulphoviridin, pigment of the anaerobe *Desulfovibrio desulfuricans*. J Gen Microbiol 14:545–572
Postgate JR (1979) The sulphate-reducing bacteria. Cambridge University Press, Cambridge
Poth M (1986) Dinitrogen production from nitrite by a *Nitrosomonas* isolate. Appl Environ Microbiol 52:957–959
Poth M, Focht DD (1985) $^{15}N$ kinetic analysis of $N_2O$ production by *Nitrosomonas europaea*: an examination nitrifier denitrification. Appl Environ Microbiol 49:1134–1141
Powell SJ, Prosser JI (1985) The effect of nitrapyrin and chloropicolinic acid on ammonium oxidation by *Nitrosomonas europaea*. FEMS Microbiol Lett 28:51–54
Prince RC, Hooper AB (1987) Resolution of the hemes of hydroxylamine oxidoreductase by redox potentiometry and electron spin resonance spectroscopy. Biochemistry 26:970–974
Probst I, Bruschi M, Pfennig N, LeGall J (1977) Cytochrome *c*-551.5($c_7$) from *Desulfuromonas acetoxidans*. Biochim Biophys Acta 460:58–64
Pronk JT, de Bruyn JG, Bos P, Kuenen JG (1992) Anaerobic growth of *Thiobacillus ferrooxidans*. Appl Environ Microbiol 58:2227–2230
Pronk JT, Liem K, Bos P, Kuenen JG (1991a) Energy transduction by anaerobic ferric iron respiration in *Thiobacillus ferrooxidans*. Appl Environ Microbiol 57:2063–2068
Pronk JT, Meijer WM, Hazeu W, van Dijken JP, Bos P, Kuenen JG (1991b) Growth of *Thiobacillus ferrooxidans* on formic acid. Appl Environ Microbiol 57;2057–2062
Puustinen A, Wikström M (1991) The heme groups of cytochrome *o* from *E. coli*. Proc Natl Acad Sci USA 88:6122–6126
Quastel JH, Scholefield PG, Stevenson JW (1952) Oxidation of pyruvic oxime by soil organisms. Biochem J 51:278–284
Quentmeier A, Kraft R, Kostka S, Klockenkamper R, Friedrich CG (2000) Characterization of a new type of sulfite dehydrogenase from *Paracoccus pantotrophus* GB17. Arch Microbiol 175:117–125
Qureshi MH, Kato C, Horikoshi K (1998) Purification of *ccb*-type quinol oxidase specifically induced in a deep-sea barophilic, *Shewanella* sp. strain DB-172F. Extremophiles 2:93–99
Qureshi MH, Yumoto I, Fujiwara T, Fukumori Y, Yamanaka T (1990) A novel *aco*-type cytochrome *c* oxidase from a facultative alkalophilic *Bacillus*: purification, and some molecular and enzymatic features. J Biochem 107:480–485
Ragsdale SW, Clark JE, Ljungdahl LL, Drake HL (1983a) Properties of purified carbon monoxide dehydrogenase from *Clostridium thermoaceticum*, a nickel, iron-sulfur protein. J Biol Chem 258:2364–2369
Ragsdale SW, Lindahl PA, Münck E (1987) Mössbauer, EPR, and optical studies of the corrinoid/iron-sulfur protein involved in the synthesis of acetyl coenzyme A by *Clostridium thermoaceticum*. J Biol Chem 262:14289–14297
Ragsdale SW, Ljungdahl LG, Der Vartanian DV (1983b) Isolation of carbon monoxide dehydrogenase from *Acetobacterium woodii* and comparison of its properties with those of the *Clostridium thermoaceticum* enzyme. J Bacteriol 155:1224–1237
Rappen H, von Berg R, Hinkel I, Thoene B, Rennenberg H (1989) Heterotrophic nitrification by *Alcaligenes faecalis*: $NO_2^-$, $NO_3^-$, $N_2O$, and NO production in exponentially growing cultures. Appl Environ Microbiol 55:2068–2072
Reis VM, Baldani JI, Baldani VLD, Dobereiner J (2000) Biological nitrogen fixation in gramineae and palm trees. Crit Rev Plant Sci 19:227–247
Richter O-MH, Tao J, Turba A, Ludwig B (1994) A cytochrome $ba_3$ functions as a quinol oxidase in *Paracoccus denitrificans*. Purification, cloning, and sequence comparison. J Biol Chem 269:23079–23086
Ritchie GAF, Nicholas DJD (1972) Identification of sources of nitrous oxide produced by oxidative and reductive processes in *Nitrosomonas europaea*. Biochem J 126:1189–1191
Ritchie GAF, Nicholas DJD (1974) The partial characterization of purified nitrite reductase and hydroxylamine oxidase from *Nitrosomonas europaea*. Biochem J 138:471–480

Robertson LA, Van Niel EWJ, Torremans RAM, Kuenen JG (1988) Simultaneous nitrification and denitrification in aerobic chemostat cultures of *Thiosphaera pantotropha*. Appl Environ Microbiol 54:2812–2818

Ronk M, Shively JE, Shute EA, Blake RC Jr (1991) Amino acid sequence of the blue copper protein rusticyanin from *Thiobacillus ferrooxidans*. Biochemistry 30:9435–9442

Rothfus JA, Smith EL (1965) Amino acid sequence of rhesus monkey heart cytochrome *c*. J Biol Chem 240:4277–4283

Rouviére PE, Wolfe RS (1988) Novel biochemistry of methanogenesis. J Biol Chem 263:7913–7916

Roy R, Knowles R (1995) Differential inhibition by allylsulfide of nitrification and methane oxidation in freshwater sediment. Appl Environ Microbiol 61:4278–4283

Saiki K, Mogi T, Anraku Y (1992) Heme O biosynthesis in *Escherichia coli*: the cyoE gene in the cytochrome *bo* operon encodes a protoheme IX farnesyl transferase. Biochem Biophys Res Commun 189:1491–1497

Sakamoto J, Hayakawa A, Uehara T, Noguchi S, Sone N (1999) Cloning of *Bacillus stearothermophilus* ctaA and heme A synthesis with CtaA protein produced in *Escherichia coli*. Biosci Biotechnol Biochem 63:96–103

Santer M, Boyer J, Santer U (1959) *Thiobacillus novellus* 1. Growth on organic and inorganic media. J Bacteriol 78:197–202

Santos H, Fareleira P, Xavier AV, Chen L, Liu M-Y, LeGall J (1993) Aerobic metabolism of carbon reserves by the "obligate anaerobe" *Desulfovibrio gigas*. Biochem Biophys Res Commun 195:551–557

Saraste M, Castresana J (1994) Cytochrome oxidase evolved by tinkering with denitrification enzyme. FEBS Lett 341:1–4

Sato A, Fukumori Y, Yano T, Kai M, Yamanaka T (1989) *Thiobacillus ferrooxidans* cytochrome *c*-552:purification and some molecular features. Biochim Biophys Acta 976:129–134

Saul RL, Kabir SH, Cohen Z, Bruce WR, Archer M (1981) Reevaluation of nitrate and nitrite levels in the human intestine. Cancer Res 41:2280–2283

Sawhney V, Nicholas DJD (1978) Sulphide-linked nitrite reductase from *Thiobacillus denitrificans* with cytochrome oxidase activity: purification and properties. J Gen Microbiol 106:119–128

Sayavedra-Soto LA, Hommes NG, Arp DJ (1994) Characterization of the gene encoding hydroxylamine oxidoreductase in *Nitrosomonas europaea*. J Bacteriol 176:504–510

Schedel M, Trüper HG (1980) Anaerobic oxidation of thiosulfate and elemental sulfur in *Thiobacillus denitrificans*. Arch Microbiol 124:205–210

Schmid M, Twachtmann U, Klein M, Juretscko S, Jetten M, Metzger JW, Schleifer K-H, Wagner M (2000) Molecular evidence for gene level diversity of bacteria capable of catalyzing anaerobic ammonium oxidation. Syst Appl Microbiol 23:93–106

Schmidt I, Bock E (1997) Anaerobic ammonia oxidation with nitrogen dioxide by *Nitrosomonas eutropha*. Arch Microbiol 167:106–111

Schmidt I, Bock E (1998) Anaerobic ammonia oxidation by cell-free extracts of *Nitrosomonas eutropha*. Antonie van Leeuwenhoek 73:271–278

Schmidt TM, DiSpirito AA (1990) Spectral characterization of c-type cytochromes purified from *Beggiatoa alba*. Arch Microbiol 154:453–458

Schröder I, Kröger A, Nacy JM (1988) Isolation of the sulphur reductase and reconstitution of the sulphur respiration of *Wolinella succinogenes*. Arch Microbiol 149:572–579

Servent D, Delaforge M, Ducrocq C, Mansuy D, Lenfant M (1989) Nitric oxide formation during microsomal hepatic denitration of glyceryl trinitrate: involvement of cytochrome P-450. Biochem Biophys Res Commun 163:1210–116

Shapiro S, Wolfe RS (1980) Methyl-coenzyme M, an intermediate in methanogenic dissimilation of $C_1$ compounds by *Methanosarcina barkeri*. J Bacteriol 141:728–734

Shethna YI, DerVartanian DV, Beinert H (1968) Non heme (iron-sulfur) proteins of *Azotobacter vinelandii*. Biochem Biophys Res Commun 31:682–688

Shiba H, Kawasumi T, Igarashi Y, Kodama T, Minoda Y (1985) The $CO_2$ assimilation via the reductive tricarboxylic acid cycle in an obligately autotrophic, aerobic hydrogen-oxidizing bacterium *Hydrogenobacter thermophilus*. Arch Microbiol 141:198–203

Shioiri M (1943) Soil chemistry of rice paddies. Data recommended for improving fertilization (in Japanese). Japan Agricultural Society

Shiratori T, Sonta H (1993) Application of iron oxidizing bacteria to hydrometallurgical flue dust treatment and $H_2S$ desulfurization. FEMS Microbiol Rev 11:165–174

Shoji K (1992) Studies on *Thiobacillus novellus* cytochrome *c* oxidase. Dissertation for Ph.D. degree, Tokyo Institute of Technology, Tokyo

Shoji K, Tanigawa M, Hori K, Tomozawa Y, Yamanaka T (1999) The effects of several nucleotides on the molecular state and catalytic activity of *Thiobacillus novellus* cytochrome *c* oxidase. Eur J Biochem 264:960–964

Shoji K, Yamazaki T, Nagano T, Fukumori Y, Yamanaka T (1992) *Thiobacillus novellus* cytochrome *c* oxidase contains one heme *a* and one copper atom per catalytic unit. J Biochem 111:46–53

Shrestha NK, Hadano S, Kamachi T, Okumura I (2001) Conversion of ammonia to dinitrogen in waste water by *Nitrosomonas europaea*. Appl Biochem Biotechnol 90:221–232

Sidloi-Lumbroso R, Schulman HM (1977) Purification and properties of soybean leghemoglobin messenger RNA. Biochim Biophys Acta 476:295–302

Silver WS, Postgate JR (1973) Evolution of asymbiotic nitrogen fixation. J Theor Biol 40:1–10

Silverman MP, Lundgren DG (1959) Studies on chemoautotrophic iron bacterium I. An improved medium and harvesting procedure for securing high cell yields. J Bacteriol 77:642–647

Silvestrini MC, Galeotti CL, Gervais M, Schinia E, Barra D, Bossa F, Burnori M (1989) Nitrite reductase from *Pseudomonas aeruginosa*: sequence of the gene and the protein. FEBS Lett 254:33–38

Silvestrini MC, Tordi MG, Citro G, Vecchini P, Brunori M (1995) Monomeric *Pseudomonas aeruginosa* nitrite reductase: preparation, characterization, and kinetic properties. J Inorg Biochem 57:169–181

Sokol WF, Evans DH, Niki K, Yagi T (1980) Reversible voltammetric response for a molecule containing four non-equivalent redox sites with application to cytochrome $c_3$ of *Dsulfovibrio vulgaris*, strain Miyazaki. J Electroanal Chem 108:107–115

Sone N (1986) Measurement of proton pump activity of the thermophilic bacterium PS3 and *Nitrobacter agilis* at the cytochrome oxidase level using total membrane and heptyl thiogucoside. J Biochem 100:1465–1470

Sone N, Yanagita Y, Hon-nami K, Fukumori Y, Yamanaka T (1983) Proton-pumping activity of *Nitrobacter agilis* and *Thermus thermophilus* cytochrome *c* oxidases. FEBS Lett 155:150–154

Sorokin DY, Kuenen JG, Jetten MS (2001) Denitrification at extremely high pH values by the alkaliphilic, obligately chemolithoautotrophic sulfur-oxidizing bacterium *Thioalkalivibrio denitrificans* strain ALJD. Arch Microbiol 175:94–101

Spieck E, Ehrich S, Aamand J, Bock E (1998) Isolation and immunocytochemical location of the nitrite oxidizing system in *Nitrospira moscoviensis*. Arch Microbiol 169:225–230

Stetter KO (1994) The lesson of archaebacteria. In: Bengston S (ed) Early life on Earth. Columbia University Press, New York, pp 143–151

Steuber J, Cypionk H, Kroneck PMH (1994) Mechanism of dissimilatory sulfite reduction by *Desulfovibrio desulfuricans*: purification of membrane-bound sulfite reductase and coupling with cytochrome $c_3$ and hydrogenase. Arch Microbiol 162:255–260

Stille W, Trüper HG (1984) Adenylylsulfate reductase in some new sulfate reducing bacteria. Arch Microbiol 137:145–150

Stolz JF, Oremland RS (1999) Bacterial respiration of arsenic and selenium. FEMS Microbiol Rev 23:615–627

Straub K, Buchholz-Cleven BE (2001) *Geobacter bremensis* sp. nov. and *Geobacter pelophilus* sp. nov., two dissimilatory ferric-iron-reducing bacteria. Int J Syst Evol Microbiol 51:1805–1808

Strauss G, Fuchs G (1993) Enzymes of a novel autotrophic $CO_2$ fixation pathway in the photosynthetic bacterium *Chloroflexus aurantiacum*, the 3-hydroxypropionate cycle. Eur J Biochem 215:633–643

Strohl WR (1989) *Beggiatoales*. In: Stanley JT, Bryant MP, Pfennig N, Holt JG (eds) Bergey's manual of systematic bacteriology vol 3. Williams & Wilkins, Baltimore, pp 2089–2106

Strous M, Fuerst JA, Kramer EH, Logemann S, Muyzer G, van de Pas-Schoonen KT, Webb R, Kuenen JG, Jetten MS (1999) Missing lithotroph identified as new planctomycetes. Nature 400:446–449

Stupperich E, Hammel KE, Fuchs G, Thauer RK (1983) Carbon monoxide fixation into the carboxyl group of acetyl coenzyme A during autotrophic growth of *Methanobacterium*. FEBS Lett 152:21–23

Stüven R, Vollmer M, Bock E (1992) The impact of organic matter on nitrite oxide formation by *Nitrosomonas europaea*. Arch Microbiol 158:439–443

Sugio T, Domatsu C, Munakata O, Tano T, Imai K (1985) Role of ferric ion-reducing system in sulfur oxidation of *Thiobacillus ferrooxidans*. Appl Environ Microbiol 49:1401–1406

Sugio T, Hirayama K, Inagaki K, Tanaka H, Tano T (1992a) Molybdenum oxidation by *Thiobacillus ferrooxidans*. Appl Environ Microbiol 58:1768–1771

Sugio T, Hirose T, Li-Zhen Y, Tano T (1992b) Purification and some properties of sulfite:ferric oxidoreductase from *Thiobacillus ferrooxidans*. J Bacteriol 174:4189–4192

Sugio T, Iwahori K, Takeuchi F, Negishi A, Maeda T, Kamimura K (2001) Cytochrome *c* oxidase purified from a mercury-resistant strain of *Acidithiobacillus ferrooxidans* volatilizes mercury. J Biosci Bioeng 92:44–49

Sugio T, Katagiri T, Inagaki K, Tano T (1989) Actual substrate for elemental sulfur oxidoreduction by sulfur: ferric oxidoreductase purified from *Thiobacillus ferrooxidans*. Biochim Biophys Acta 973:250–256

Sugio T, Kishimoto K, Takai M, Oda K, Tano T (1995) Growth of moderately thermophilic iron-oxidizing bacterium strain TI-1 in synthetic medium. J Ferment Bioeng 79:290–293

Sugio T, Kuwano H, Negishi A, Maeda T, Takeuchi F, Kamimura K (2001) Mechanism of growth inhibition by tungsten in *Acidithiobacillus ferrooxidans*. Biosci Biotechnol Biochem 65:555–562

Sugio T, Mizunashi W, Inagaki K, Tano T (1987) Purification and some properties of sulfur: ferric ion oxidoreductase from *Thiobacillus ferrooxidans*. J Bacteriol 169:4916–4922

Sugio T, Suzuki H, Ota A, Inagaki K, Tanaka H, Tano T (1991) Purification and some properties of hydrogen sulfide-binding protein that is involved in sulfur oxidation of *Thiobacillus ferrooxidans*. Agric Biol Chem 55:2091–2097

Suharti, Strampraad MJ, Schröder I, de Vries S (2001) A novel copper A containing menaquinol NO reductase from *Bacillus azotoformans*. Biochemistry 40:2632–2639

Sumino T, Isaka K, Ikuta H, Saiki Y, Yokota T (2006) Nitrogen removal from waste-water using simultaneous nitrate reduction and anaerobic ammonium oxidation in single reactor. J Biosci Bioeng 102:346–351

Sumino T, Nakamura H, Mori N, Yamaguchi Y, Tada M (1992) Immobilization of nitrifying bacteria in porous pellets of urethane gel for removal of ammonium nitrogen from waste-water. Appl Microbiol Biotechnol 36:556–560

Sundermeyer-Klinger H, Meyer W, Warninghof B, Bock E (1984) Membrane-bound nitrite oxidoreductase of *Nitrobacter*: evidence for a nitrate reductase system. Arch Microbiol 140: 153–158

Suzuki I, Chan CW, Takeuchi TL (1992) Oxidation of elemental sulfur to sulfite by *Thiobacillus thiooxidans* cells. Appl Environ Microbiol 58:3767–3769

Suzuki I, Dular U, Kwok S-C (1974) Ammonia or ammonium ion as substrate for oxidation by *Nitrosomonas europaea* cells and extracts. J Bacteriol 120:556–558

Suzuki I, Kwok S-C (1970) Cell-free ammonia oxidation by *Nitrosomonas europaea*: effects of polyamines, $Mg^{2+}$ and albumin. Biochem Biophys Res Commun 39:950–955

Suzuki I, Kwok S-C (1981) A potential resolution and reconstitution of the ammonia oxidizing system of *Nitrosomonas europaea*. Role of cytochrome *c*-554. Can J Biochem 59:484–488

Suzuki I, Kwok S-C, Dular U, Tsang DCY (1981) Cell-free ammonia-oxidizing system of *Nitrosomonas europaea*: general conditions and properties. Can J Biochem 59:477–483

Suzuki M, Cui ZJ, Ishii M, Igarashi Y (2001) Nitrate respiratory metabolism in an obligately autotrophic hydrogen-oxidizing bacterium, *Hydrogenobacter thermophilus* TK-6. Arch Microbiol 175:75–78

Suzuki Y, Fujiwara T, Fukumori Y (1997) Structure of *Nitrobacter winogradskyi* nitrite oxidoreductase. Seikagaku 69:892

Svensson B, Lübben M, Hederstedt L (1993) *Bacillus subtilis* ctaA and ctaB function in heme A biosynthesis. Mol Microbiol 10:193–201

Takai M, Kamimura M, Sugio T (1999) Involvement of cytochrome *a* in iron oxidation of moderately thermophilic iron-oxidizing bacterium, strain TI-1. Biosci Biotechnol Biochem 63:1541–1547

Takai M, Kamimura K, Sugio T (2001) A new iron oxidase from moderately thermophilic iron-oxidizing bacterium strain TI-1. Eur J Biochem 268:1653–1658

Takakuwa S (1976) Studies on the metabolism of a sulfur-oxidizing bacterium XI. Electron transfer and the terminal oxidase system in sulfite oxidation of *Thiobacillus thiooxidans*. Plant Cell Physiol 17:103–110

Tal Y, Watts JE, Schreier HJ (2005) Anaerobic ammonia-oxidizing bacteria and related activity in Baltimore inner harbor sediment. Appl Environ Microbiol 71:1816–1821

Tamegai H, Fukumori Y (1994) Purification, and some molecular and enzymatic features of a novel *ccb*-type cytochrome *c* oxidase from microaerobic denitrifier, *Magnetospirillum magnetotacticum*. FEBS Lett 347:22–26

Tamegai H, Kai M, Fukumori Y, Yamanaka T (1994) Two membrane-bound *c*-type cytochromes of *Thiobacillus ferrooxidans*: purification and properties. FEMS Microbiol Lett 119:145–154

Tanaka Y, Fukumori Y, Yamanaka T (1982) The complete amino acid sequence of *Nitrobacter agilis* cytochrome *c*-550. Biochim Biophys Acta 707:14–20

Tanaka Y, Fukumori Y, Yamanaka T (1983) Purification of cytochrome $a_1c_1$ from *Nitrobacter agilis* and characterization of nitrite oxidase system of the bacterium. Arch Microbiol 135:265–271

Tano T, Ito T, Takesue H, Sugio T, Imai K (1982) B-type cytochrome, an electron carrier in the sulfite-oxidation system of *Thiobacillus thiooxidans*. J Ferment Technol 60:181–187

Taya M, Shiraishi H, Katsunishi T, Tone S (1991) Enhanced cell density culture of *Thiobacillus ferrooxidans* in membrane-type bioreactor with electrolytic reduction unit for ferric ion. J Chem Eng Jpn 24:291–296

Taylor CD, Wolfe RS (1974) Structure and methylation of coenzyme M ($HSCH_2$–$CH_2SO_3^-$). J Biol Chem 249:4879–4885

Tedro SM, Meyer TE, Kamen MD (1979) Primary structure of a high potential, four-iron-sulfur ferredoxin from the photosynthetic bacterium, *Rhodospirillum tenue*. J Biol Chem 254:1495–1500

Tedro SM, Meyer TE, Kamen MD (1985) The amino acid sequence of a high redox-potential ferredoxin from the phototrophic bacterium, *Rhodospirillum tenue*. Arch Biochem Biophys 239:94–101

Thauer RK, Bonacker LG (1994) Biosynthesis of coenzyme $F_{430}$, a nickel porphinoid involved in methanogenesis. In: The biosynthesis of the tetrapyrrole pigments (Ciba Foundation Symposium 180). Wiley, Chichester, pp 210–222

Thode HG (1980) Sulphur isotope ratios in late and early precambrian sediments and their implications regarding early environments and early life. Origins Life 10:127–136

Timkovich R, Bergmann D, Arciero DM, Hooper AB (1998) Primary sequence and solution conformation of ferrocytochrome *c*-552 from *Nitrosomonas europaea*. Biophys J 75:1964–1972

Timkovich R, Cork MS, Gennis RB, Johnson PY (1985) Proposed structure of heme *d*, a prosthetic group of bacterial terminal oxidase. J Am Chem Soc 107:6069–6075

Toghrol F, Southerland WM (1983) Purification of *Thiobacillus novellus* sulfite oxidase. Evidence for the presence of heme and molybdenum. J Biol Chem 258:6762–6766

Tomizuka N, Takahara Y (1972) Bacterial leaching of uranium from Ningyo-Toge ores. In: Terui G (ed) Fermentation technology today. Society of Fermentation Technology, Osaka, pp 513–520

Trudinger PA (1961a) Thiosulphate oxidation and cytochromes in *Thiobacillus* X. 1. Fractionation of bacterial extracts and properties of cytochromes. Biochem J 78:673–680

Trudinger PA (1961b) Thiosulphate oxidation and cytochromes in *Thiobacillus* X. 2. Thiosulphate-oxidizing enzyme. Biochem J 78:680–686

Tsang DCY, Suzuki I (1982) Cytochrome *c*-554 as possible electron donor in hydroxylation of ammonia and carbon monoxide in *Nitrosomonas europaea*. Can J Biochem 60:1018–1024

Tsukihaha T, Aoyama H, Yamashita E, Tomizaki T, Yamaguchi H, Shinzawa-Itoh K, Nakashima R, Yaono R, Yoshikawa S (1995) Structures of metal sites of oxidized bovine heart cytochrome *c* oxidase at 2.8 Å. Science 269:1069–1072

Tsukihara T, Aoyama H, Yamashita E, Tomizuka T, Yamaguchi H, Shinzawa-Itoh K, Nakashima R, Yaono R, Yoshikawa S (1996) The whole structure of the 13-subunit oxidized cytochrome *c* oxidase at 2.8 Å. Science 272:1136–1144

Uchida A (1995) Distribution and biosynthesis of dimethyl sulfide. In: Matsumoto S (ed) Microbial ecology 20: Microbial gas metabolism and Earth environments (in Japanese). Scientific Societies Press, Tokyo, pp 145–163

Uchiyama H (1999) Halogeno-organic compounds and microorganisms. In: Japanese Society of Soil Microorganisms (ed) New microorganisms in soil (4): Environments and microorganisms (in Japanese). Hakuyu-sha, Tokyo, pp 71–90

Upadhyay AK, Hooper AB, Hendrich MP (2006) NO reductase activity of the tetraheme cytochrome *c*-554 of *Nitrosomonas europaea*. J Am Chem Soc 128:4330–4337

Valkova-Valchanova MB, Chan SHP (1994) Purification and characterization of two new *c*-type cytochromes involved in $Fe^{2+}$ oxidation from *Thiobacillus ferrooxidans*. FEBS Lett 288: 159–162

Van Beelen P, Van Neck JW, de Cock RM, Vogels GD, Guijt W, Haasnoot CAG (1984) 5,10-methenyl-5,6,7,8-tetrahydromethanopterin, a one carbon carrier in the process of methanogenesis. Biochemistry 23:4448–4454

Van Beeumen JJ, Ambler RP (1973) Homologous in the amino acid sequences of cytochrome *c*-555 from the green photosynthetic bacteria *Chloropseudomonas ethylica* and *Chlorobium thiosulfatophilum*. Antonie van Leeuwenhoek 39:355–356

Van der Oost J, de Boer AP, de Gier JW, Zumft WG, Stouthamer AH, van Spanning RJ (1994) The heme–copper oxidase family consists of three distinct types of terminal oxidases and is related to NO reductase. FEMS Microbiol. Lett 121:1–10

Vaupel M, Thauer RK (1998) Two $F_{420}$-reducing hydrogenases in *Methanosarcina barkeri*. Arch Microbiol 169:201–205

Vavra MR, Timkovich R, Yap F (1986) Spectroscopic studies on heme *d* in the visible and infrared. Arch Biochem Biophys 250:461–468

Vellieux FMD, Frank J Jr, Swarte MBA, Groendijk H, Duine JA, Drenth J, Hol WGJ (1986) Purification, crystallization and preliminary X-ray investigation of quinoprotein methylamine dehydrogenase from *Thiobacillus versutus*. Eur J Biochem 154:383–386

Vernon LP, Kamen MD (1954) Haematin compounds in photosynthetic bacteria. J Biol Chem 211:643–662

Vernon LP, Mangum JH, Beck JV, Shafia FM (1960) Studies on a ferrous-iron-oxidizing bacterium II. Cytochrome composition. Arch Biochem Biophys 88:227–231

Verstraete W, Alexander M (1973) Heterotrophic nitrification in samples of natural ecosystems. Environ Sci Technol 7:39–42

Vincent S (ed) Nitric oxide in the nervous system. Academic Press, London

Visscher PT, Taylor BF (1993) A new mechanism for the aerobic catabolism of dimethyl sulfide. Appl Environ Microbiol 59:3784–3789

Visser JM, de Jong GAH, de Vries S, Robertson LA, Kuenen JG (1997a) *cbb*$_3$-type cytochrome oxidase in the obligately chemolithoautotrophic *Thiobacillus* sp. W5. FEMS Microbiol Lett 147:127–132

Visser JM, de Jong GAH, Robertson LA, Kuenen JG (1997b) Purification and characterization of a periplasmic thiosulfate dehydrogenase from the obligately autotrophic *Thiobacillus* sp. W5. Arch Microbiol 166:372–378

Visser JM, de Jong GAH, Robertson LA, Kuenen JG (1997c) A novel membrane-bound flavocytochrome *c* sulfide dehydrogenase from the colourless sulfur bacterium *Thiobacillus* sp. W5. Arch Microbiol 167:295–301

Völkl P, Huber R, Drobner E, Rachel R, Burggraf S, Trincone A, Stetter KO (1993) *Pyrobaculum aerophilum* sp. nov., a novel nitrate-reducing hyperthermophilic archaeum. Appl Environ Microbiol 59:2918–2926

Vosjan JH (1970) ATP generation by electron transport in *Desulfovibrio desulfuricans.* Antonie van Leeuwenhoek 36:584–586

Wächtershäuser G (1988) Pyrite formation, the first energy source for life: a hypothesis. Syst Appl Microbiol 10:207–210

Wakabayashi S, Matsubara H, Webster DA (1986) Primary sequence of dimeric bacterial hemoglobin from *Vitreoscilla.* Nature 322:481–483

Wakao N, Yokoi N, Isoyama N, Hiraishi A, Shimada K, Kobayashi M, Kise H, Iwaki M, Itoh S, Takaichi S, Sakurai Y (1996) Discovery of natural photosynthesis using Zn-containing bacteriochlorophyll in aerobic bacterium *Acidiphilum rubrum.* Plant Cell Physiol 37:889–893

Warren MJ, Bolt E, Woodcock SC (1994) 5-Aminolaevulinic acid synthase and uroporphyrinogen methylase: two control enzymes of tetrapyrrole biosynthesis and modification. In: Biosynthesis of the tetrapyrrole pigments (Ciba foundation symposium 180). Wiley, Chichester, pp 26–40

Watanabe H, Kamita Y, Nakamura T, Takimoto A, Yamanaka T (1979) The terminal oxidase of *Photobacterium phosphoreum.* A novel cytochrome. Biochim Biophys Acta 547:70–78

Watson SW, Bock E, Harms H, Koops H-P, Hooper AB (1989) Nitrifying bacteria. In: Stanley JT, Bryant MP, Pfennig N, Holt JG (eds) Bergey's manual of systematic bacteriology vol 3. Williams & Wilkins, Baltimore, pp 1808–1834

Wehrfritz J-M, Carter JP, Soiro S, Richardson DJ (1997) Hydroxylamine oxidation in heterotrophic nitrate-reducing soil bacteria and purification of hydroxylamine cytochrome *c* oxidoreductase from *Pseudomonas* species. Arch Microbiol 166:421–424

Wehrfritz J-M, Reilly A, Spiro S, Richardson DJ (1993) Purification of hydroxylamine oxidase from *Thiosphaera pantotropha.* Identification of electron acceptors that couple heterotrophic nitrification to aerobic denitrification. FEBS Lett 335:246–250

Wetzstein H-G, Ferguson SJ (1985) Respiration-dependent proton translocation and the mechanism of proton motive force generation in *Nitrobacter winogradskyi.* FEMS Microbiol Lett 30:87–92

Widdel F (1986) Growth of methanogenic bacteria in pure culture with 2-propanol and other alcohols as hydrogen donors. Appl Environ Microbiol 51:1056–1062

Winogradsky SN (1891) Ueber die Organismen der Nitrification. Vierteljahrsschrift der Naturforschenden Gesellschaft in Zürich 16:176–208

Wittenberg JB (1974) Facilitated oxygen diffusion. The role of leghemoglobin in nitrogen fixation by bacteroids isolated from soybean root nodules. J Biol Chem 249:4057–4066

Wittenberg JB, Appleby CA, Wittenberg BA (1972) The kinetics of the reactions of leghemoglobin with oxygen and carbon monoxide. J Biol Chem 247:527–531

Wodara C, Bardischewsky F, Friedrich CA (1997) Cloning and characterization of sulfite dehydrogenase, two *c*-type cytochromes, and a flavoprotein of *Paracoccus denitrificans* GB17: essential role of sulfite dehydrogenase in lithotrophic sulfur oxidation. J Bacteriol 179:5014–5023

Woese CR, Fox GE (1977) Phylogenetic structure of the prokaryotic domain: the primary kingdoms. Proc Natl Acad Sci USA 74:5088–5090

Wolfe RS (1980) Respiration in methanogenic bacteria. In: Knowles CJ (ed) Diversity of bacterial respiratory systems vol I. CRC Press, Boca Raton, FL, pp 161–186

Wood HG, Ljungdahl LG (1991) Autotrophic character of the acetogenic bacteria. In: Shively J-M, Barton LL (eds) Variations in autotrophic life. Academic Press, London, pp 201–250

Wood PM (1986) Nitrification as bacterial energy source. In: Prosser JL (ed) Nitrification. IRL Press, Oxford, pp 39–62

Yagi T (1979) Purification and properties of cytochrome *c*-553, an electron acceptor for formate dehydrogenase of *Desulfovibrio vulgaris* Miyazaki. Biochim Biophys Acta 548:96–105

Yagi T, Honya M, Tamiya N (1968) Purification and properties of dehydrogenases of different origins. Biochim Biophys Acta 153:699–705

Yagi T, Maruyama K (1971) Purification and properties of cytochromes $c_3$ of *Desulfovibrio vulgaris* Miyazaki. Biochim Biophys Acta 243:214–224

Yamagata S (1939) Über die Natratreduktase und die "Nitritreduktase", ein neues Enzym, von *Bacillus pyocyaneus*. Untersuchungen über die biologischen Reduktion II. Acta Phytochim 11:145–157

Yamanaka T (1964) Identification of *Pseudomonas* cytochrome oxidase with *Pseudomonas* nitrite reductase. Nature 204:253–255

Yamanaka T (1975) A comparative study on the redox reactions of cytochromes *c* with certain enzymes. J Biochem 77:493–499

Yamanaka T (1983) Effect of paraquat on growth of *Nitrosomonas europaea* and *Nitrobacter agilis*. Plant Cell Physiol 24:1394–1352

Yamanaka T (1992) The biochemistry of bacterial cytochromes. Japan Scientific Societies Press, Tokyo/Springer, Berlin

Yamanaka T (1994) Sulfide-cytochrome *c* reductase (flavocytochrome *c*) Methods Enzymol 243:463–472

Yamanaka T (1996) Mechanisms of oxidation of inorganic electron donors in autotrophic bacteria. Plant Cell Physiol 37:569–574

Yamanaka T, Aso I, Togashi S, Tanigawa M, Shoji K, Watanabe T, Watanabe N, Maki K, Suzuki H (2002a) Corrosion by bacteria of concrete in sewerage systems and inhibitory effects of formates on their growth. Wat Res 36:2636–2642

Yamanaka T, Fujii K (1980) Cytochrome *a*-type terminal oxidase derived from *Thiobacillus novellus*. Molecular and enzymatic properties. Biochim Biophys Acta 591:53–62

Yamanaka T, Fujii K, Kamita Y (1979a) Subunits of cytochrome *a*-type terminal oxidases derived from *Thiobacillus novellus* and *Nitrobacter agilis*. J Biochem 86:821–824

Yamanaka T, Fujiwara T, Yano T, Shoji K, Fukumori Y (1995) Evolutional position of chemoautotrophic bacteria based on the structure of cytochrome *c*. Viva Origino 23:179–188

Yamanaka T, Fukumori Y (1977) *Thiobacillus novellus* cytochrome oxidase can separate eukaryotic cytochromes *c*. FEBS Lett 77:155–158

Yamanaka T, Fukumori Y (1978) An evolutionary aspect of the reactivity of cytochrome *c* with cytochrome oxidase. In: Matsubara H, Yamanaka T (eds), Evolution of protein molecules. Japan Scientific Societies Press, Tokyo, pp 387–399

Yamanaka T, Fukumori Y (1980) A biochemical comparison between *Chlorobium* and *Chromatium* flavocytochromes *c*. In: Yagi K, Yamanao T (eds) Flavins and flavoproteins. Japan Scientific Societies Press, Tokyo, pp 631–639

Yamanaka T, Fukumori Y (1988) Nitrite oxidizing system of *Nitrobacter winogradskyi*. FEMS Microbiol Rev 54:259–270

Yamanaka T, Fukumori Y (1995) Molecular aspects of the electron transfer system which participate in the oxidation of ferrous ion by *Thiobacillus ferrooxidans*. FEMS Microbiol Rev 17:401–413

Yamanaka T, Fukumori Y, Yano T, Kai M, Sato A (1991a) Enzymatic mechanisms in the "dehydrogenation" of ferrous ions by *Thiobacillus ferrooxidans*. In: Berthelin J (ed) Diversity of environmental biogeochemistry. Elsevier, Amsterdam, pp 267–273

Yamanaka T, Kamita Y, Fukumori T (1981a) Molecular and Enzymatic properties of "cytochrome $aa_3$"-type terminal oxidase derived from *Nitrobacter agilis*. J Biochem 89:265–273

Yamanaka T, Kijimoto S, Okunuki K (1963) Biological significance of *Pseudomonas* cytochrome oxidase in *Pseudomonas aeruginosa*. J Biochem 53:416–421

Yamanaka T, Kimura K (1974) Eukaryotic cytochrome *c*-like properties of cytochrome *c*-550 (*Thiobacillus novellus*). FEBS Lett 48:253–255

Yamanaka T, Miyasaka H, Aso I, Tanigawa M, Shoji K, Yohta H (2002b) Involvement of sulfur- and iron-transforming bacteria in heaving of house foundations. Geomicrobiol J 19:519–528

Yamanaka T, Nagano T, Shoji K, Fukumori Y (1991b) Cytochromes *c* of *Nitrobacter winogradskyi* and *Thiobacillus novellus*: structure, function and evolution. Biochim Biophys Acta 1058:48–51

Yamanaka T, Okunuki K (1963a) Crystalline *Pseudomonas* cytochrome oxidase I. Enzymatic properties with special reference to biological specificity. Biochim Biophys Acta 67:379–393

Yamanaka T, Okunuki K (1963b) Crystalline *Pseudomonas* cytochrome oxidase II. Spectral properties of the enzyme. Biochim Biophys Acta 67:394–406

Yamanaka T, Okunuki K (1963c) Crystalline *Pseudomonas* cytochrome oxidase III. Properties of the prosthetic groups. Biochim Biophys Acta 67:407–416

Yamanaka T, Ota A, Okunuki K (1960) Nitrite reductase activity of *Pseudomonas* cytochrome oxidase. Biochim Biophys Acta 44:397–398

Yamanaka T, Ota A, Okunuki K (1961) A nitrite reducing system reconstituted with purified cytochrome components of *Pseudomonas aeruginosa*. Biochim Biophys Acta 53:294–308

Yamanaka T, Sakano Y (1980) Oxidation of hydroxylamine to nitrite catalyzed by hydroxylamine oxidoreductase purified from *Nitrosomonas europaea*. Curr Microbiol 4:239–244

Yamanaka T, Shinra M (1974) Cytochrome *c*-552 and cytochrome *c*-554 derived from *Nitrosomonas europaea*: purification, properties and their function. J Biochem 75:1265–1273

Yamanaka T, Shinra M, Takahashi K, Shibasaka M (1979b) Highly purified hydroxylamine oxidoreductase derived from *Nitrosomonas europaea*. J Biochem 86:1101–1108

Yamanaka T, Takenami S, Akiyama N, Okunuki K (1971) Purification and properties of cytochrome *c*-550 and cytochrome *c*-551 derived from the facultative chemoautotroph, *Thiobacillus novellus*. J Biochem 70:349–358

Yamanaka T, Tanaka Y, Fukumori Y (1982) *Nitrobacter agilis* cytochrome *c*-550:isolation, physicochemical and enzymatic properties, and primary structure. Plant Cell Physiol 23:441–449

Yamanaka T, Yoshioka T, Kimura K (1981b) Purification of sulphite-cytochrome *c* reductase of *Thiobacillus novellus* and reconstitution of its sulphite oxidase system with the purified constituents. Plant Cell Physiol 22:613–622

Yamazaki Takehiro, Oyanagi H, Fujiwara T, Fukumori Y (1995) Nitrite reductase from the magnetotactic bacterium *Magnetospirillum magnetotacticum*. A novel cytochrome $cd_1$ with Fe(II): nitrite oxidoreductase activity. Eur J Biochem 233:665–671

Yamazaki Takeshi, Fukumori Y, Yamanaka T (1985) Cytochrome $a_1$ of *Nitrosomonas europaea* resembles $aa_3$-type cytochrome *c* oxidase in many respects. Biochim Biophys Acta 810: 174–183

Yamazaki Takeshi, Fukumori Y, Yamanaka T (1988) Catalytic properties of cytochrome *c* oxidase purified from *Nitrosomonas europaea*. J Biochem 103:499–503

Yano T (1992) The oxidation system of ferrous ion in *Thiobacillus ferrooxidans*. Dissertation for Ph.D. degree, Tokyo Institute of Technology, Tokyo

Yano T, Fukumori Y, Yamanaka T (1991) The amino acid sequence of rusticyanin isolated from *Thiobacillus ferrooxidans*. FEBS Lett 288:159–162

Yaoi Y (1967) Comparison of the primary structures of cytochromes *c* from wild and respiration-deficient mutant yeasts. J Biochem 61:54–58

Yarzabal A, Brasseur G, Bonnefoy V (2002a) Cytochromes of *Acidithiobacillus ferrooxidans*. FEMS Microbiol Lett 209, 189–195

Yarzabal A, Brasseur G, Ratouchniak J, Lund K, Lemesle-Meunier D, DeMoss JA, Bonnefoy V (2002b) The high molecular weight cytochrome *c* Cyc2 of *Acidithiobacillus ferrooxidans* is an outer membrane protein. J Bacteriol 184:313–317

Yohta H (2000) A study on heaving due to biochemical weathering of the Neogene sedimentary soft rock. Dissertation for Ph.D. degree, College of Science and Technology, Nihon University, Tokyo

Yoon K-S, Ishii M, Kodama T, Igarashi Y (1997) Carboxylation reactions of pyruvate: ferredoxin oxidoreductase from *Hydrogenobacter thermophilus* TK-6. Biosci Biotechnol Biochem 61:510–513

Yoshida N (1988) $^{15}N$-depleted $N_2O$ as a product of nitrification. Nature 335:528–529

Yoshino T (1994) Studies on nitrite-cytochrome *c* oxidoreductase (cytochrome $a_1c_1$) of *Nitrobacter winogradskyi.* Dissertation for Master Degree, Tokyo Institute of Technology, Tokyo

Zahn JA, Arciero DM, Hooper AB, DiSpirito AA (1996) Evidence for an iron center in the ammonia monooxygenase from *Nitrosomonas europaea.* FEBS Lett 397:35–38

Zahn JA, Duncan C, DiSpirito AA (1994) Oxidation of hydroxylamine by cytochrome P-460 of the obligate methylotroph *Methylococcus capsulatus* Bath. J Bacteriol 176:5879–5887

Zart D, Schmidt I, Bocke E (2000) Significance of gaseous NO for ammonia oxidation by *Nitrosomonas europaea.* Antonie van Leeuwenhoek 77:49–55

Zeikus JG (1977) The biology of methanogenic bacteria. Bacteriol Rev 41:514–541

Zimmermann BH, Nitsche CI, Fee JA, Rusnack F, Munk E (1988) Properties of a copper-containing cytochrome $ba_3$:a second terminal oxidase from the extreme thermophile *Thermus thermophilus.* Proc Natl Acad Sci USA 85:5779–5782

Zimmermann P, Laska S, Kletzin A (1999) Two modes of sulfite oxidation in the extremely thermophilic and acidophilic archaeon *Acidianus ambivalens.* Arch Microbiol 172:76–82

Zumft WG, Braun C, Cuypers H (1994) NO reductase from *Pseudomonas stutzer.* Primary structure of gene organization of a novel bacterial *bc* complex. Eur J Biochem 219:481–490

Zumft WG, Matsubara T (1982) A novel kind of multicopper protein as terminal oxidoreductase of $N_2O$ respiration in *Pseudomonas perfectomarinus.* FEBS Lett 148:107–112

# Index


**a**

acetate 113
acetyl phosphate 54–56
acetyl-CoA 112–113, 117
acetyl-CoA synthesizing pathways 109, 116
acetylene 49
*Achromobacter* sp 38
*Acidianus ambivalens* 62, 66, 68
*Acidianus brierleyi* 66, 82, 109
*Acidianus infernus* 120
acidic rain 17
*Acidithiobacillus acidophilus* 67
*Acidithiobacillus ferrooxidans* 8, 26, 69–70, 79–84, 86–90, 94–95, 100
  cytochrome *c* (46 kDa) 82, 87
  cytochrome $c_4$ 82–83, 85–86
  cytochrome *c*-550(m) (51 kDa) 82–83, 85–86
  cytochrome *c*-552(m) (22.3 kDa) 82–83, 85–86
  cytochrome *c*-552(s) (14 kDa) 81–83, 85–87
  cytochrome *c*-553(s) (12789 Da) 82
*Acidithiobacillus thiooxidans* 8, 65, 67, 71–72, 76
acidophilic iron-oxidizing bacteria 76–77
adenosine 5′-diphosphate (ADP) 1
adenosine 5′-triphosphate (ATP) 1
adenylylsulfate (APS) 54–55, 58
adenylylsulfate reductase 58, 68
ADP 56
*akiochi* 62
*Alcaligenes faecalis* 38
*Alcaligenes faecalis* strain TUD 37
alcohol fermentation 4
algae 53, 103
*Allochromatium* (formerly *Chromatium*) *vinosum* 62, 64
ammonia monooxygenase 19, 27, 37
ammonia-oxidizing bacteria 5, 18, 115
AMP 56
anaerobic chemolithoautotrophs 74
anammox bacteria 40
angina pectoris 46–47
animal hemoglobin 50
antibacterial agent 77
APS 56, 58
archaebacteria (archaea) 119–120
arsenic respiration 61
ascorbate peroxidase 43
assimilatory sulfite reductase 58–59
ATP 55, 56
ATP synthase 87, 112
*Azospirillum* genus 50–51
*Azotobacter vinelandii* 15, 48, 50
azurin 45

**b**

*Bacillus azotoformans* 46
bacterial leaching 94–95
Bacterium S 112
*Beggiatoa alba* 66
*Beggiatoa* genus 8, 72
biohydrometallurgy 97
bioleaching 94
bioremediation 30

**c**

$C_1$ cycle 110
$C_3$-plants 107–108
$C_4$ dicarboxylate pathway 105
$C_4$-plants 107–108
calcium formate 77–78
Calvin–Benson cycle 105–107

carbon dioxide 103–105
carbon dioxide respiration 112, 124
carbon monoxide 104, 116
carbon monoxide dehydrogenase (CODH) 109, 116–117
catalase 43
chalcopyrite 94
chemical evolution 122
chemoheterotrophic bacteria 2
chemoheterotrophic organisms 104
chemolithoautotrophs 105–106, 121
chemolithoautotrophic bacteria 4, 103
*Chlorobium ferrooxidans* 80, 91
*Chlorobium limicola* 64
*Chlorobium limicola* f. *thiosulfatophilum* 62
*Chlorobium limicola* f. *thiosulfatophilum* cytochrome *c*-555 25
*Chloroflexus aurantiacus* 108
*Citrobacter freundii* 91
*Clostridium thermoaceticum* 116
CODH 109, 116–117
coenzyme $F_{420}$ 112–113
coenzyme $F_{430}$ 59, 110, 112–113
coenzyme M 110
copper 94
copper protein-type nitrite reductase 45
coronary artery of heart 46
corrinoid enzyme 109, 117
corrosion of concrete 74–78
cyanobacteria 51, 53, 92, 103, 123
cyclic guanosine 3′,5′-monophosphate (cGMP) 47
cytochrome *a* 14, 88
cytochrome $a_1$ 25–26, 32, 84
cytochrome $a_1a$ 26
cytochrome $a_1c_1$ 14, 26, 32, 34
cytochrome $a_3$ 14, 69
cytochrome $aa_3$ 14, 69, 84–85
cytochrome *aco* 14, 16
cytochrome *b* 14, 61–62, 70, 71, 114
cytochrome $ba_3$ 14
cytochrome $baa_3$ 14
cytochrome *bc* 114
cytochrome $bc_1$ 13, 88
cytochrome *bd* 16, 50, 85
cytochrome $bo_3$ 16
cytochrome *c* 15, 40, 41, 61, 71
cytochrome *c* oxidase 13, 14, 25, 34, 68–69, 81, 83, 86, 88, 116
cytochrome $c_2$ 15
cytochrome $c_3$ 15, 56, 57, 59, 61, 114
cytochrome $c_3$ (26 kDa) 57
cytochrome $c_4$ 15
cytochrome $c_5$ 15
cytochrome *c*-551.5 62
cytochrome $c_6$ 15, 25, 57
cytochrome $c_7$ 62
cytochrome $c_8$ 15, 25
cytochrome $cao_3$ 14
cytochrome *cbb* 45
cytochrome $cbb_3$ 15, 26, 69
cytochrome $cd_1$ 15, 45, 93
cytochrome $cd_1$-type nitrite reductase 45
cytochrome *co* 16
cytochrome P-450 46
cytochrome P-460 21–22
cytochromes 11, 13, 124

**d**
*Dechlorosoma suillum* 7, 92
*Dehalococcoides ethenogenes* 28
dehalogenation of tetrachloroethylene 29
dehalogenation respiration 29
denitrification 44
denitrifying bacteria 44, 47
*Desulfotomaculum* genus 55
*Desulfovibrio auripigmentum* 61
*Desulfovibrio desulfuricans* 54, 57–58
*Desulfovibrio* genus 57
cytochrome *c*-553 56–57
*Desulfovibrio gigas* 55–57, 61
*Desulfovibrio vulgaris* 7, 61–62
*Desulfuromonas acetoxidans* 61–62
dichloroethylene 29
dihydrosirohydrochlorin 59
dimethyl sulfide 53
dinitrogen tetraoxide 31
dissimilatory sulfite reductase 58

**e**
electrotroph 89
elemental sulfur 53, 123
*Escherichia coli* 58
eubacteria 119
eukarya 119
eukaryotes 119

**f**
$F_0F_1$-ATPase 35–36, 87
$F_1$-type ATPase 36
facultative chemolithoautotrophic bacteria 8
Fe(II)-cytochrome *c* oxidoreductase 81, 83, 85–87
fermentation 3, 112
ferredoxin 56
ferredoxin-NAD(P) reductase 43
ferric hydroxide 91–92
ferric iron reductase 92

*Ferroglobus placidus* 7, 80, 91, 120
ferrous rusticyanin oxidoreductase 83
flavocytochrome *c* 62, 66
formate dehydrogenase 57

**g**
*Gallionella ferruginea* 9, 79–80
*Geobacter genus* 9
*Geobacter metallireducens* 92
*Geobacter sulfurreducens* 92
*Geoglobus ahangari* 93
*Geospirillum* 91
glucose 121–122
glycolysis 121
gold 96–97
green sulfur bacteria 123
greenhouse effect 103, 109, 115
guanosine 5′-triphosphate (GTP) 47
guanyl cyclase 47–48
gunpowder 39
gypsum 102

**h**
$H_2$:heterodisulfide oxidoreductase 110, 114–115
*Halobacterium salinarum* 120
Hatch–Slack pathway 105, 107
heme A 12
heme B 12
heme C 12
heme D 12
heme $D_1$ 12
heme O 12
heme P-460 15, 21
hemoglobin 11
hemoglobin of tube worm 73
herbicide 43
heterocysts 51
high-potential iron–sulfur protein [HiPIP] 81
HOQNO 70, 89–90
human tissues 41
hydrogen cyanide 49
hydrogen sulfide 53, 54, 75, 77
hydrogenases 56, 116
*Hydrogenobacter thermophilus* 109, 119, 123
hydrothermal vents 61, 72–73, 97, 121
hydroxylamine oxidase system 28
hydroxylamine oxidoreductase 15, 19–20, 22–24, 27
3-hydroxypropionate cycle 108
hyperthermophilic bacteria 123
*Hyphomicrobium* 53

**i**
inhibitors of ammonia monooxygenase 20
iron oxidase of cytochrome *a*-type 85
iron oxidizing bacteria 8, 9, 79, 82
iron respiration 124
iron-oxidizing bacteria 85, 95–101
iron-reducing bacteria 93
isotopic ratio of sulfur 63

**j**
jarosite 102

**k**
Kuroko 97–98

**l**
lactate dehydrogenase 57
lactic acid 116
leghemoglobin 49–50
*Leptospirillum ferrooxidans* 79–80, 82, 84, 90
*Leptothrix discophora* 90

**m**
magnetite 92–93
magnetosome 93
*Magnetospirillum magnetotacticum* 93
magnetotactic bacteria 93
*Metallosphaera sedula* 82
methane 63, 104, 109–110, 113, 115
methane-oxidizing bacteria 110
*Methanobacterium bryantii* 112
*Methanobacterium* genus 114
*Methanobacterium omelianskii* 112
*Methanobacterium thermoautotrophicum* 109, 115
*Methanobacterium thermoautotrophicum* ΔH 10
*Methanococcus igneus* 120
*Methanococcus jannaschii* 120
*Methanococcus voltae* 110
methanogens 9, 10, 104, 109–110, 112, 115–116, 119, 124
*Methanosarcina barkeri* 109, 114
*Methanosarcina* sp. 113
*Methanothrix thermophila* 113
methemoglobin 47
methyl-CoM 113
methyl-CoM reductase 110, 113
methyl viologen 42
*Methylococcus capsulatus* Bath 22

*Methylococcus thermophilus* 37
*Methylophilus methylotrophus* 16
*Microcoleus chthonoplastes* 66
mine sewage 98
molecular oxygen 92
molybdenum 48
molybdenum cofactors 45
mudstone 100–101

**n**

NAD(P) reductase 43
NAD(P)H-cytochrome *c* oxidoreductase 35
NADH dehydrogenase-1 88
*nakabosi* 62
*Natronomonas pharaonis* 120
nickel 110, 113, 116
Ningyo-rock 95
niter 39
nitrate 46–47
nitrate reductase 45
nitrate respiration 3, 44–45
nitric oxide (NO) 30, 35, 42, 44, 46–48
nitric oxide reductase (NO reductase) 13, 24, 45–46
nitric oxide synthase (NOS) 47
nitrification-controlling reagents 20
nitrite oxidoreductase 13, 32–33
nitrite reductase 13, 15, 24–25, 45
nitrite-oxidizing bacteria 6
*Nitrobacter hamburgensis* 31–33
*Nitrobacter winogradskyi* 5, 26, 31, 33, 35–36, 41–43
  cytochrome $a_1c_1$ 35
  cytochrome $aa_3$ 34–35
  cytochrome *b*-559 35
  cytochrome *c*-550(m) 34–35
  cytochrome *c*550(s) 32, 34–35, 40
nitrogen dioxide 30–31, 41–42
nitrogen fixation 49
nitrogenase 48, 50–51
nitrogenase complex 48
nitrogen-fixing bacteria 47
nitroglycerin 46
*Nitrosococcus oceanus* 18
*Nitrosomonas europaea* 5, 18, 24–25, 28, 30–31, 37, 39, 42–43
  cytochrome $aa_3$ 26
  cytochrome *b*-560 27
  cytochrome *c*-552 25, 27–28, 40
  cytochrome *c*-554 19, 23–25, 27–28
  cytochrome *c* oxidase 26
*Nitrosomonas eutropha* 18, 30–31
*Nitrospira moscoviensis* 32–33
nitrous oxide ($N_2O$) 23–24, 39, 44
nitrous oxide reductase ($N_2O$ reductase) 46

**o**

oleic acid 28
origin of life 63–64
oxidation of manganese 91
oxyhemoglobin 47
ozone layer 39

**p**

*Paracoccus denitrificans* 14, 19, 39
*Paracoccus denitrificans* GB17 67
*Paracoccus pantotropha* GB17 37, 39
*Paracoccus versutus* 67, 69, 71
paraquat 42–43
PEP-carboxylase 107
perchloroethylene 29
phospholipids 119
*Photobacterium phosphoreum* 16
photosynthesis 103, 123
photosynthetic sulfur bacteria 64
pitchblende 95
poly-L-lysine 26
protoheme 12
proton pumping activity of cytochrome *c* oxidase 34, 36, 69
*Pseudomonas aeruginosa* 16, 45
  cytochrome *c*-551 25, 45
*Pseudomonas aeruginosa* nitrite reductase 25, 45
*Pseudomonas mendocina* cytochrome *c*-551 25
*Pseudomonas* PB16 37
*Pseudomonas* S-36 91
*Pseudomonas* sp. HD-1 10
*Pseudomonas stutzeri* 16
  cytochrome *c*-552 25
pyrite ($FeS_2$) 53, 87, 96, 100, 121
*Pyrobaculum aerophilum* 72
*Pyrobaculum islandicum* 92–93, 120, 123–124
*Pyrococcus furiosus* 120
pyruvate synthase [pyruvate-ferredoxin 2-oxidoreductase (CoA-acetylating)] 57
pyruvic oxime 38
pyruvic oxime dioxygenase 38

**q**

quinol oxidase 13, 70

**r**
reconstitution of a nitrite oxidase system 35
reductive carboxylate cycle 109
reductive pentose phosphate cycle 105
relaxation of smooth muscles 47
respiration 112
rhizobia 50
rhodanese 66
*Rhodobacter sphaeroides* 26
*Rhodopseudomonas marina* 66
root nodules 49–50
Rubisco 105
ruminants 115
rusticyanin 81–84, 87

**s**
$^{32}S/^{34}S$ 64
S.N. Winogradsky 4
selenium respiration 61
Shethna protein 50
silver 96–97
siroheme 58–59
sirohydrochlorin 59
*Starkeya novella* 8, 65–67, 69, 71
  cytochrome *c*-550 67–68, 70
*Starkeya novella* cytochrome *c* oxidase 26, 70
structures of hemes 12
sulfate 53, 123
sulfate adenylyltransferase 55, 68
sulfate adenylyltransferase (ADP) 68
sulfite cytochrome *c* oxidoreductase 71
sulfite dehydrogenase 67
sulfite reductase 57–59
sulfate respiration 3, 7, 54, 55, 114, 124
sulfite-acceptor oxidoreductase 68
sulfite-cytochrome *c* oxidoreductase 67–68, 70
sulfate-reducing bacteria 54–55, 62–63, 74, 99, 101, 115
sulfide-cytochrome *c* oxidoreductase 65
sulfide-Fe(III) oxidoreductase 88–89
*Sulfolobus acidocaldarius* 26, 120
sulfur dioxide 53
sulfur dioxygenase 66, 71
sulfur reductase 61–62
sulfur respiration 7, 61, 124
sulfur respirers 61
sulfur-accepting protein 66, 71
sulfuric acid 54
sulfur-oxidizing bacteria 8, 64–65, 73–77, 99, 101
superoxide anion 43
superoxide dismutase (SOD) 43

**t**
tetrachloroethylene 28
tetrathionate 72
tetrathionate hydrolase 67
*Thioalkalivibrio denitrificans* 72
*Thiobacillus caldus* 71
*Thiobacillus denitrificans* 6, 72
*Thiobacillus neapolitanus* 8, 65, 67, 72, 76
*Thiobacillus* sp. W5 66–67, 69
*Thiobacillus tepidarius* 67, 70–72
*Thiobacillus thioparus* 67
*Thioploca* genus 72
*Thiosphaera pantotropha* 39
thiosulfate 72
thiosulfate reductase 57
thiosulfate-cleaving enzyme 66, 71
thiosulfate-cytochrome *c* oxidoreductase 67
thiosulfate-sulfur transferase 66
*Thiovulum* genus 73
trichloroethylene 28–29
tube worms 73

**u**
ubiquinone-8 27–28
upheaval of house foundations 99, 101
uranium 95
uroporphyrinogen III 59–60, 113

**v**
vanadium 48
vinyl chloride 29
vitamin $B_{12}$ 59, 112
vitamin $B_{12}$-containing enzyme (corrinoid enzyme) 109, 117

**w**
withering of vegetables 41
*Wolinella succinogenes* 62

**y**
yeast 41